MW00378591

TOYS FROM OCCUPIED JAPAN

With Price Guide

Anthony R. Marsella

Schiffer Publishing Ltd

77 Lower Valley Road, Atglen, PA 19310

DEDICATION

Dedicated to two of the most important people in my life.

To my friend, fellow toy collector, and partner, Bernard G. Toll (Boinie!), who owns a great number of the toys in this book and whose friendship and guidance have been a motivational factor in my life.

To my mother, Mrs. Nora Marsella (Mommie!), who, at her yard sale, told me "Save those two items, Anthony. They are marked 'Occupied Japan' and I understand that they are becoming more and more valuable."

The utmost of thanks to the both of you.

Copyright 1995 by Anthony Marsella

Library of Congress Catalog Card Number: 95-69835

All rights reserved. No part of this work may be reproduced or used in any forms or by any means—graphic, electronic, or mechanical, including photocopying or information storage and retrieval systems—without written permission from the copyright holder.

Printed in Hong Kong
ISBN: 0-88740-875-3

Published by Schiffer Publishing Ltd.
77 Lower Valley Road
Atglen, PA 19310
Please write for a free catalog.
This book may be purchased from the publisher.
Please include $2.95 for shipping.
Try your bookstore first.

We are interested in hearing from authors with book ideas on related subjects.

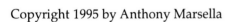

CONTENTS

ACKNOWLEDGMENTS

A book, like a movie, is not possible without a good supporting staff during its production. A special word of thanks to the following for their understanding and support.

...to Peter and Nancy Schiffer for their patience during my chronically late deadlines. I realize that this book would probably never have come close to being finished without their understanding of my illness and their recognition that the time was right for this book.

...to Douglas Congdon-Martin, my editor, for the same amount of patience.

...to Bernard G. Toll, my friend and fellow collector, for his assistance in gathering up his collection to lend me for this book.

...to Mikki Deatherage, Curt and Lynette Parmer, Frank Travis, and Margaret Bolbat, fellow O.J. collectors, for their confidence and support.

Thank you! Gracie!! Grazia!!!

INTRODUCTION

"Why don't you keep those two figurines, Anthony? They are marked 'Occupied Japan'. I understand that any items marked as such are becoming more valuable and collectible." My mother said these words one fateful morning in 1971, as we were preparing to have a yard sale. She told me that she had recently read that because of the short time span of the occupation, anything made in Japan during those years and exported to the United States had to be marked accordingly.

Needless to say, she really aroused my curiosity on the subject and within a few years I was picking up just about anything of interest marked Occupied Japan, or as collectors like to say, O.J.

For about the next ten years I became an avid collector, picking up pieces at garage sales and flea markets. By the early to mid 1980s I had accumulated a thousand or so items of all varieties from figurines to cigarette lighters, notions to religious items, full-size sewing machines to "TOYS". My interest, as with most collectors, began to specialize in what I think is the most intriguing middle class collectible of our time.

To help finance this hobby I decided to sell off all the O.J. in the collection other than toys. I wanted to immediately reinvest the money, so I began attending organized toy shows, looking furiously for that O.J. mark on the toy, box, or, whenever pos-sible, both. For the O.J. purist, this is the only way to authenticate its originality.

The historical definition of O.J. has been told in many books to date, but here is a brief synopsis.

From September, 1945 until April, 1954, the army of the Occupation from the United States of America set up its headquarters in Tokyo under General Douglas MacArthur. The foremost order to the army was to aid in the restoration of the economy of the country. Though many of the factories had been destroyed during the war, there were still enough operable ones to resume manufacturing. Being the industrious people that they always were, the Japanese set up one of the world's strongest work forces and by the end of the Occupation had an economy already stronger than their conqueror's.

A very innovative people, far ahead of their time in thought, they recycled tin from cans to make toys. Imagine my surprise when I look inside the "Justa Shmoe" car in this book and found that it had been recycled from a "Schmidt's Beer" can. The irony was that this beer was made in Philadelphia, Pennsylvania...U.S.A....and exported to Japan.

For a more detailed history of Occupied Japan, may I suggest *Occupied Japan* for Collectors by Florence Archambault. It is also published by Schiffer Publishing, Ltd.

Part One Mechanical Toys

CHAPTER ONE
ANIMAL TOYS

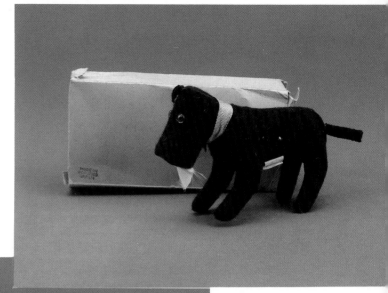

Dog with handkerchief. 5" x 2.5".

ANIMALS

"'B' Dog." 5" x 4".

"Wire Dog." 3.5" x 4.5".

"Playful Poodle." 3" x 5".

"Jumping Dog." 3" x 4".

"Begging Poodle." 2" x 5".

Gray dog. 3" x 4".

Black Dachshund with bone. 3" x 6".

Brown furry poodle. 3" x 4".

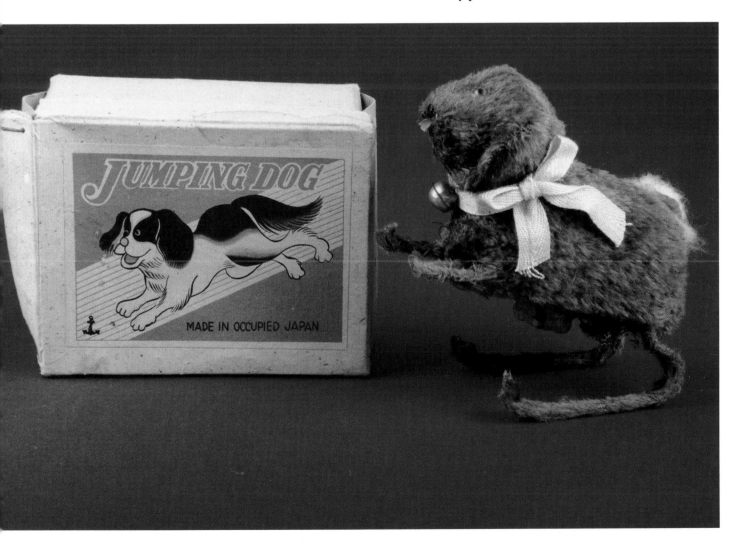

"Jumping Dog" with peg legs. 2" x 4".

"Roll Over Cat." 3" x 5".

Gold fur cat. 3.5" x 4".

"Polar Bears." 3" x 5".

"Walking Bear." 3" x 5".

"Walking Bear." 6" x 3".

"Walking Small Bear." 4" x 5.5"

13

Baby Bear. 2" x 3".

"Skating Bear." 4" x 7".

"Fishing Bear." 6" x 3".

"Tita" Monkey. 5" x 5".

"Mr. Joe, the Educated Monkey." 3" x 5".

15

"New Style Circus Monkey." 6" x 5".

"Wandering Hairy Chimpanzee." 3" x 6". Y.

"Banjo Player" monkey. 2" x 5".

Gray furry monkey. 4" x 12".

Furry monkey. 2" x 6".

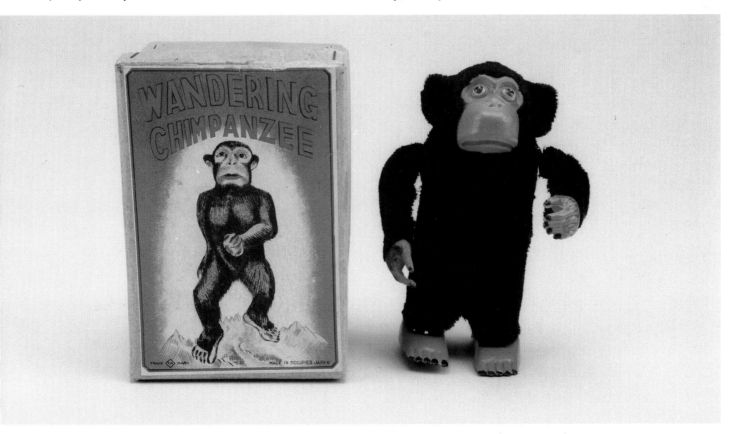

"Wandering Chimpanzee." 2" x 5.5".

"Elephant." 4" x 6".

"Circus Elephant." 3" x 6".

"Baby Elephant." 3" x 4".

18

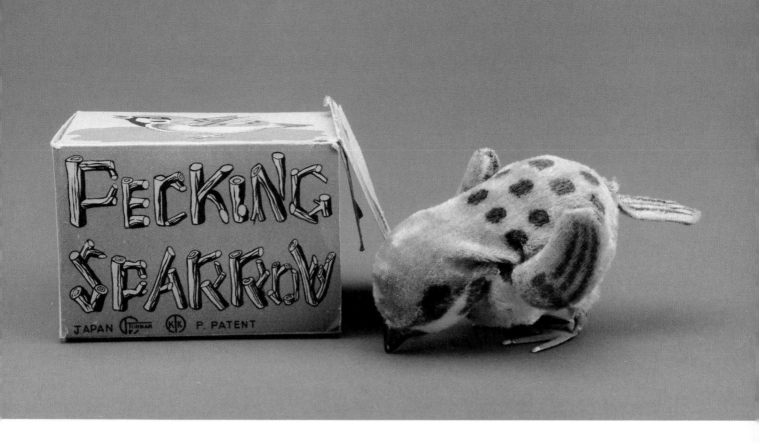

"Pecking Sparrow." 5" x 2".

"Chuck Chick." 2" x 3".

19

"Quacy Wacky." 5" x 5".

"Skating Ostrich." 4" x 6".

Penguin with
pipe. 2" x 4.5".

"Chirping Bird." 3" x 6".

20

"Dandy Master." 2" x 6".

Furry rabbit. 4" x 4".

"Friendly Rabbit." 3" x 4".

Begging rabbit. 3" x 5".

"Roaring Lion." 4" x 7".

Lion. 3" x 6".

"Walking Goat," with lead horns. 3.5" x 5".

"Sea Lion." 3" x 4".

"Mother & Baby Kangaroo." 8" x 6.5".

Spider. 2" x 3".

Squirrel, "Skipper in the Forest." 3" x 6".

"Nutty Squirrel." 5" x 2".

"Travelling Camel." 4" x 6".

Donkey. 3" x 4".

"Camel." 4" x 5".

L: Tin giraffe; R: "Plush Giraffe" with original box. 4" x 7".

Plush & Tin "Yearling." Same box went with both toys. 4" x 6".

"Tumbling Fido." 3" x 5"

"Playful Little Dog." 3" x 4" Y.

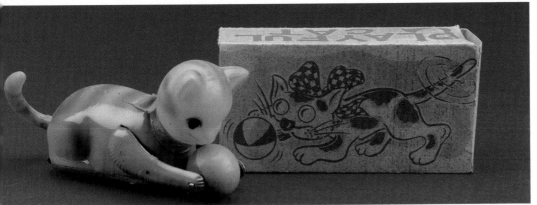

"Playful Cat," tin with celluloid head. 3" x 6".

Eating monkey. 2" x 4".

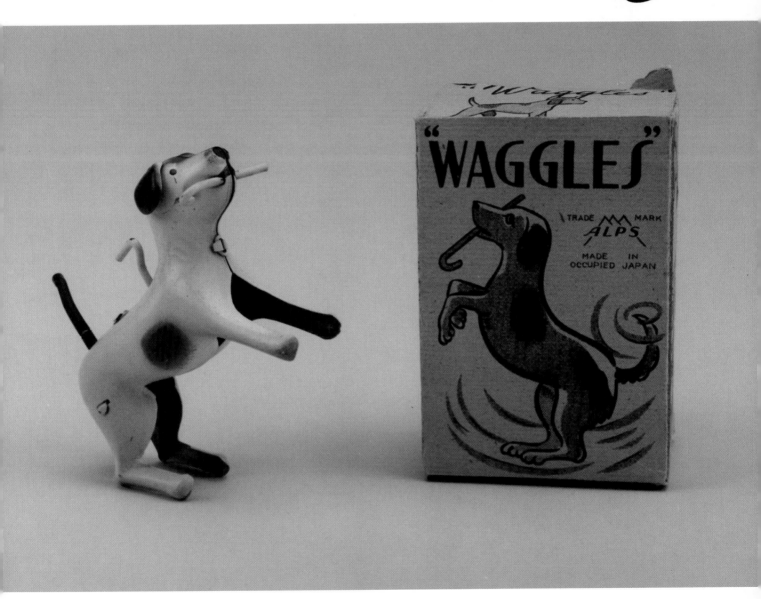

"Waggles" the dog. 2." x 4".

"Circus Elephant." 3" x 6"

Elephant. 3" x 4".

"Jitterbug Elephant." 3" x 4".

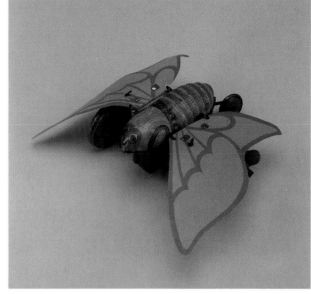

"Walking 'Beetle'." 1.5" x 4".   Butterfly. 1" x 4".

"Proud Peacock." 6" x 6".

"Walking Duck." 3" x 3".

"Pick-Chick." 2" x 2".

Tin penguin. 1.5" x 3".

"Penguin," tin. 1" x 3".

Duck, tin. 2" x 3".

"Mechanical Egg Laying Ducks" with celluloid eggs. The one on the left is lithographed tin, while the one on the right is hand painted. 3" x 4".

Tin bouncing chick. 2" x 3".

"Singing Chicken." 4" x 4".

"Swan" with baby. 4" x 6.5".

"Pig's Joker." 1.5" x 3".

Rabbit, tin. 5" x 2".

"Sea Lion." 3" x 4".

Turtle. 1" x 4".

"Tortoise". 1" x 4".

"Crocodile." 1" x 6".

"Kangaroo." 2" x 3.5".

36

"Tumbling Mouse." 3" x 3".

"Frog." 1" x 3".

Gray Mice. 1.5" x 3".

CELLULOID ANIMALS

"Walkie Doggie." 4" x 3".

"Come, come, Dog." 2" x 4".

Scotty with hat. 2" x 4".

"Playing Dog." 2" x 4".

Puppy with bee. 1" x 3".

"Short Stop Dog." 2" x 4".

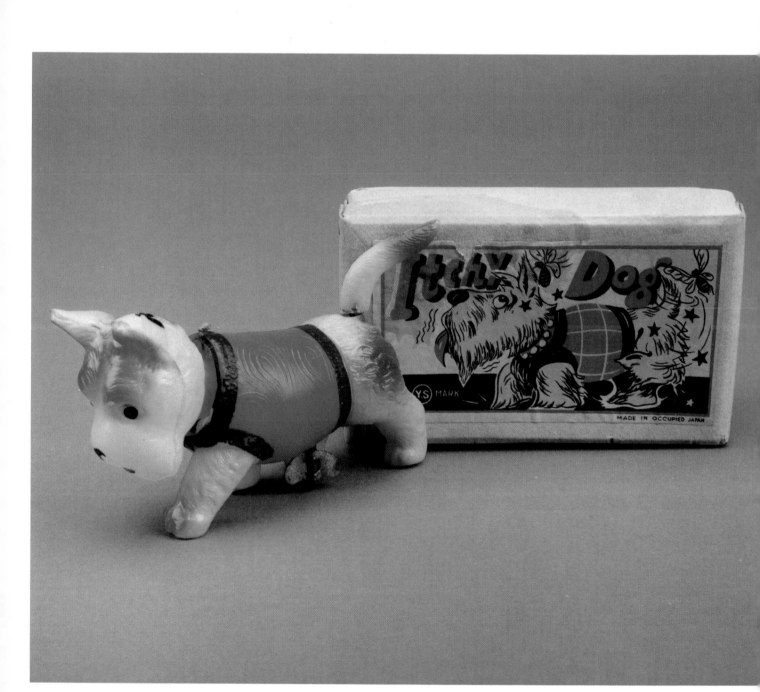

"Itchy Dog." 3" x 4".

"Puppy & Bee." 3" x 4".

"Shoes Dog." 2" x 6".

"Itchy Dog." 3" x 4".

Dog with cat. 3" x 6".

"Kennel Frolics." 2" x 9".

Vaudeville trick dog and rolling hat. 4" x 6".

"Cat Rolling Ball." 4" x 3".

"Cat Butterfly Chaser." 2" x 5".

Cat with ball on belly. 3" x 4".

"Kitty Ball Romp." 3" x 7".

Cat bouncing ball. 2" x 4.5"

"Sarcas" monkey. 2" x 6".

"Gymnastic Monkey." 5" x 7".

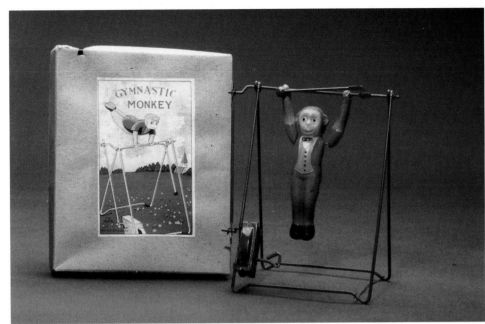

Tumbling monkey. 3" x 3".

"Rollo The Monk." 5" dia.

45

"Monkey Cycle." 2" x 3".

Tumbling monkey. 2" x 3".

"Jim Dandy." 2" x 5".

"'Sambo' The Minstrel Man." 3.5" x 8".

"Monkey Sweet Melodian." 2" x 4".

"Tumbling Jocko," monkey. 4" x 3".

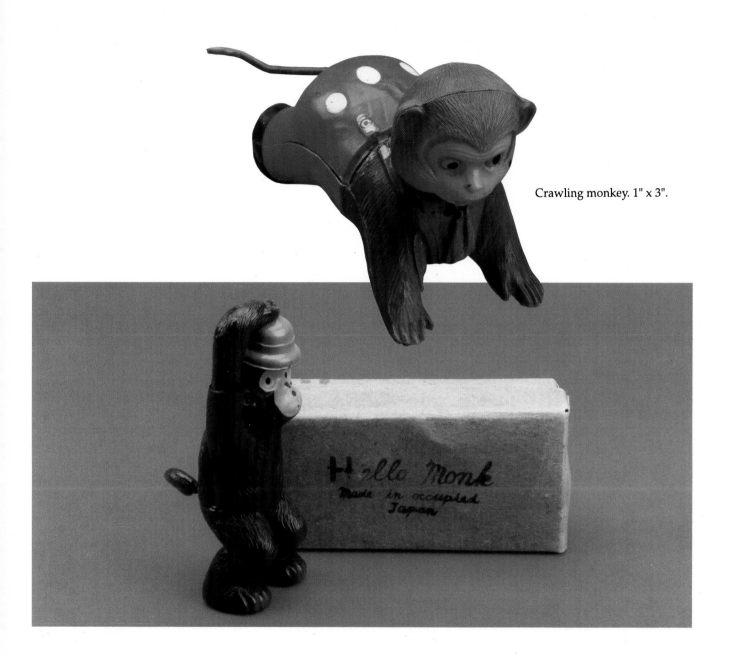

Crawling monkey. 1" x 3".

"Hello Monk." 2" x 4".

Celluloid squirrel with bushy tail. 4" x 6".

"Sarcas" elephant. 2" x 6".

"Moving Elephant." 3" x 5".

49

Jumbo-type elephant. 2" x 4".

"Elephant on Barrel." 3" x 8".

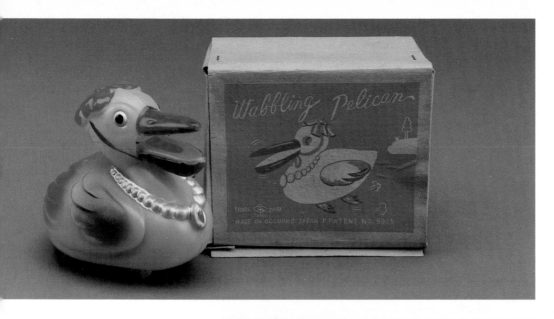

"Wobbling Pelican." 4" x 3.5".

"Strolling Duck." 2" x 5".

"Strolling Pelican." 5.5" x 4"

Celluloid swan. 2" x 4".

"Petey the Penguin." 2" x 5".

"Duck & Frog." 2" x 5".

"Singing Baby Chicks." 2" x 4".

"Singing Canary." 2" x 3".

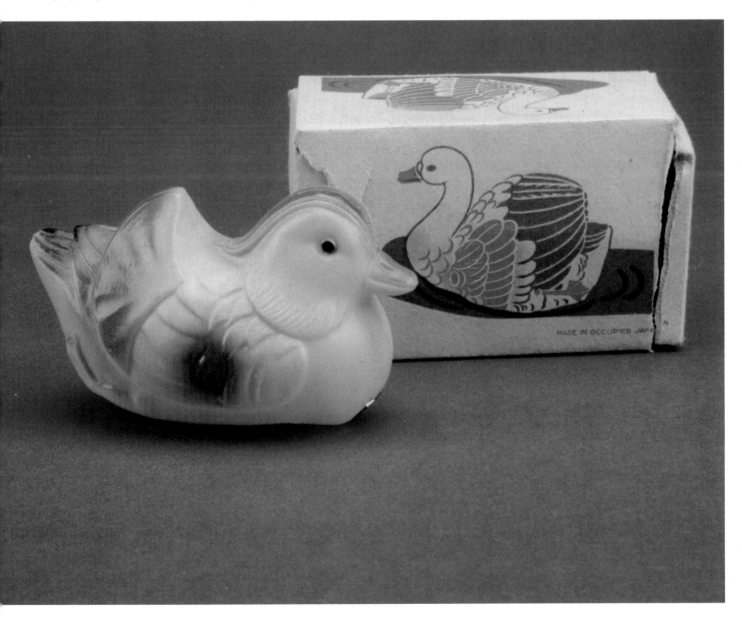

Swan. 2.5" x 4".

Tumbling celluloid bear. 2" x 4".

"Circus Bear." 2" x 7".

"Proud Bear." 2" x 4".

54

"Roaring Lion." 3" x 5".

"Lion Teaser." 8" x 5".

"Bleating Pig." 5" x 2.5".

"Gentleman Frog." 2" x 4".

"Crawling Tortoise." 2" x 5".

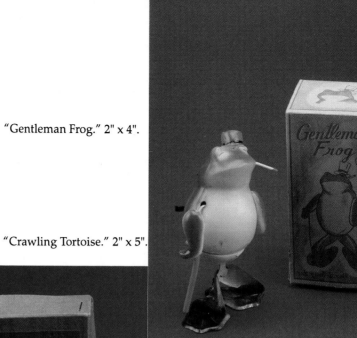

"Trick Seals," in the two color variations. 4" x 5".

L: Bunny on tricycle, 4" X 5"; R: pig on tricycle, 4" x 5".

"Banjo Bunny," came in two different colors. 3.5" x 8".

57

CARS

Large blue sedan. 2" x 6".

CHAPTER TWO TRANSPORTATION

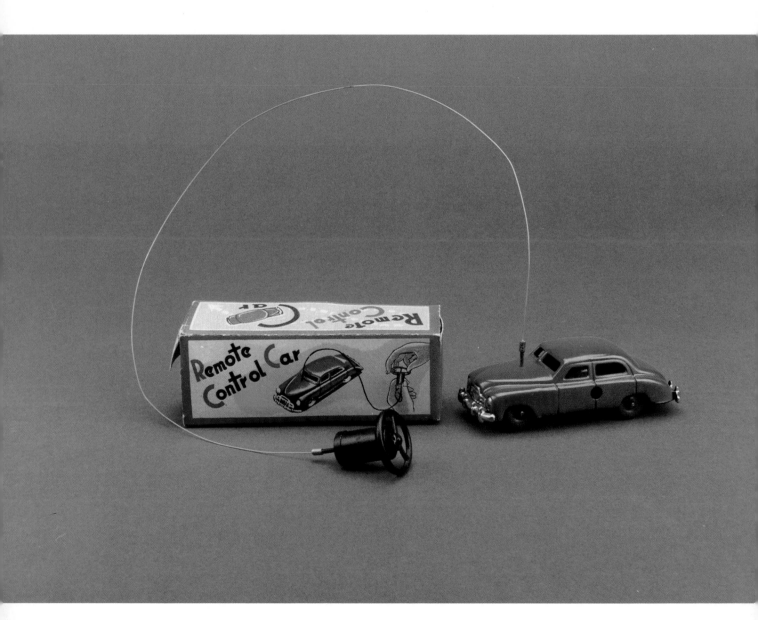

"Remote Control Car." 1.5" x 4".

"Baby Pontiacs." 1" x 3.5".

"New Buick." 2" x 4".

"Friction Car." 1.5" x 4".

"X-Car." 2" x 4".

"Chevrolet" with back motion. 2" x 4".

Military car. 2" x 4".

"Signal Car." 3" x 5".

"Studebaker." 2" x 4".

"Puzzle Car." 2.5" x 6".

"Lucky Car." 2" x 5".

Pink and red Hudson. 3" x 6".

Hudson. 3" x 7".

"New Friendly Car." 3" x 6".

Blue coupe (Chevy). 3" x 5".

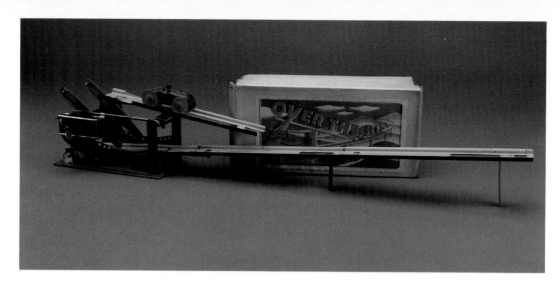

"Over The Hill." 5" x 22".

"Hurricane Racer." 2" x 5".

"Race car." 3" x 5".

"Baby Race Car." 1.5" x 3".

"Justa Shmoe" car, made from a Schmidt's beer can. 4.5" x 3".

Flip-over racer. 3" x 6".

"Baby Halloo Jeep." 1" x 2.5".

"Fire Engine." 2" x 4"

"Sparkling G-Man Car." 3" x 7".

"Police Car." 2" x 6".

Black G-man auto. 3" x 6".

"G-Men Car." 2" x 6".

BUSES AND TROLLEYS

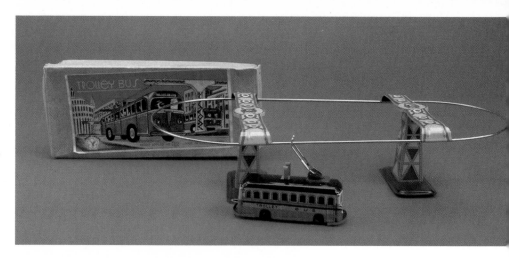

"Trolley Bus." 8" x 15".

Bus. 1.5" x 4".

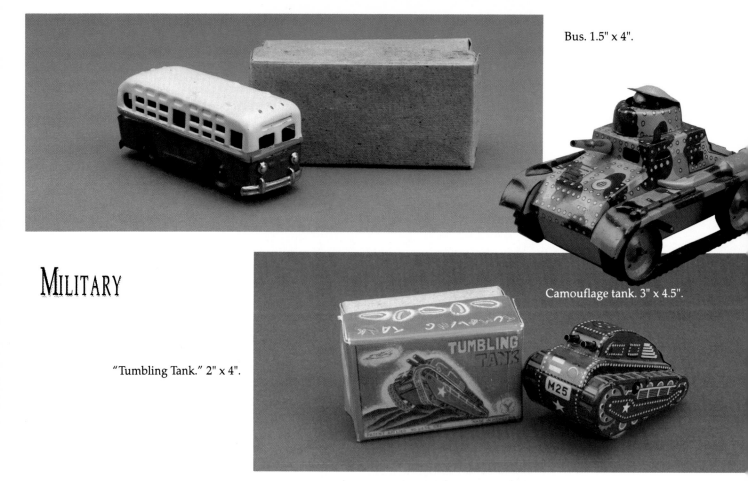

MILITARY

Camouflage tank. 3" x 4.5".

"Tumbling Tank." 2" x 4".

"Sparkling Tank." 1.5" x 3".

Puzzle Jeep. 4" x 6".

Green Jeep. 5" x 3".

CONSTRUCTION

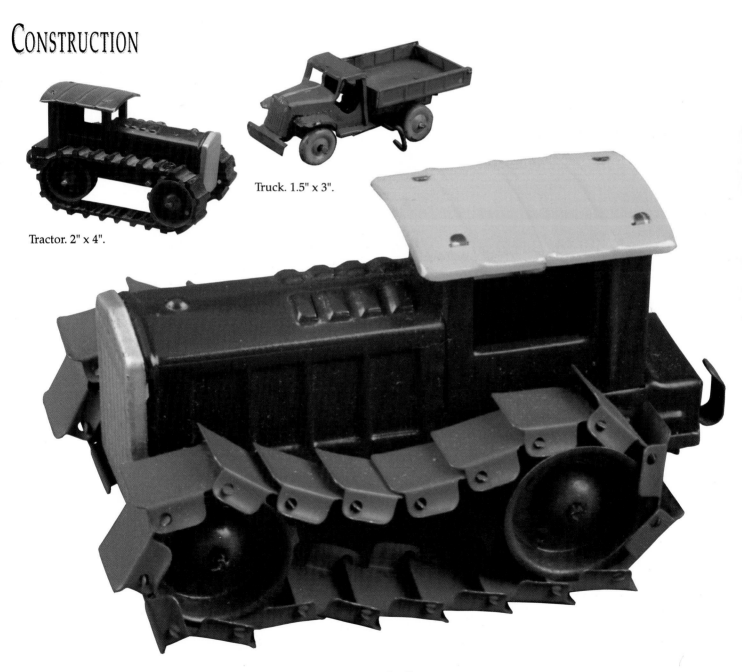

Tractor. 2" x 4".

Truck. 1.5" x 3".

Tractor. 2" x 4".

"Harley."

"Auto Cycle," hand-painted driver. 3" x 5".

"Auto Cycle," lithographed driver. 3" x 5".

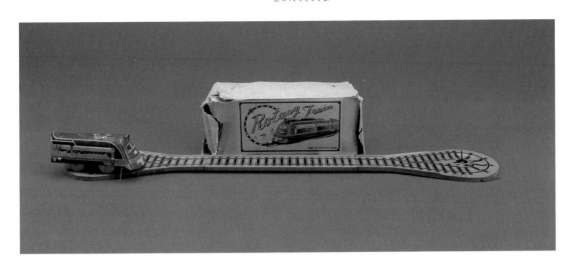

"Rotary Train." 17" x 2".

Train engine. 1" x 4".

"Shunting Train." 2.5" x 12".

"Electric Engine." 3.5" x 9".

"Speedy Electric Car." 3" x 10".

"Cyclic Train." 5" x 10".

Airplanes

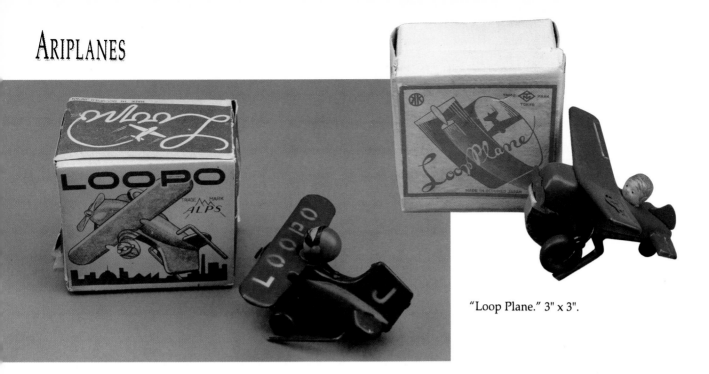

"Loop Plane." 3" x 3".

"Loopo" plane. 3" x 3".

Gyrocopter. 2" x 6".

"Sparkling Loopo Plane." 2" x 3.5".

"Pleasure Trip Plane." 5" x 7" (wingspan).

BOATS

Battery operated boat. 5" x 16".

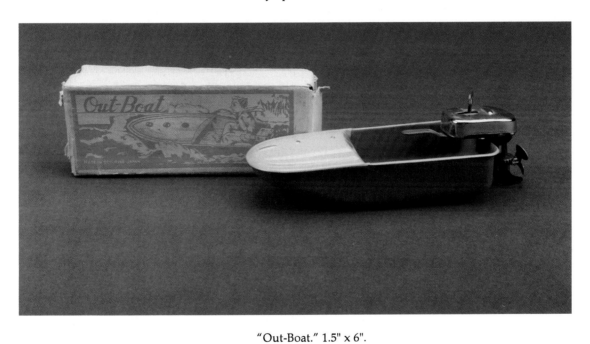

"Out-Boat." 1.5" x 6".

"Out Board Motor," battery operated. 2" x 6".

CHAPTER THREE PEOPLE

"Cow Boy with Two Guns." 2" x 5".

Celluloid "Cowboy" with lasso. 1" x 3".

COWBOYS AND INDIANS

Celluloid cowboy with lasso. 1" x 3".

"Cow-Boy." 5" x 5".

Celluloid cowboy on tin horse. 4" x 4".

"Cowboy." 6" x 7".

"Indian." 6" x 7".

"Indian on Horse." 5" x 5".

"Gay Caballero." 5" x 5".

"Sancho Panza." 5" x 5".

Tin cowboy with lasso.
1.5" x 4".

"Rancher." 4" x 6".

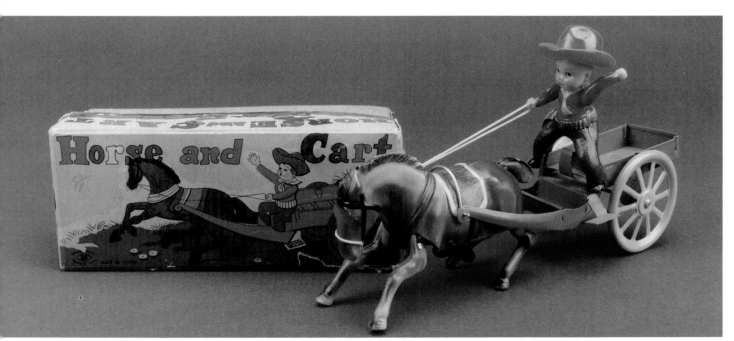

"Horse & Cart." 4" x 10".

"Prairie Schooner." 4" x 9".

"Prairie Wagon." 4" x 9".

Boy with donkey cart. 4" x 8".

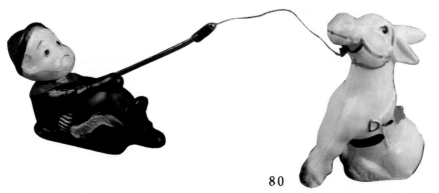

Cowboy with donkey. 3" x 5".

"Shimmy Donkey." 3" x 4".

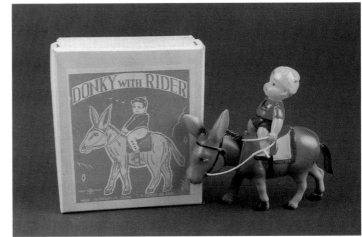

Boy on donkey. 4" x 5".

"Bucking Jeep." All lithographed tin. 4" x 5".

COUPLES

"Dancing Couple." 2" x 4.5".

Small "Dancing Couple." 2" x 3".

Larger "Dancing Couple." 2" x 4.5".

Small sailor "Dancing Couple." 2" x 3".

Skating couple with hat." 3" x 5".

"Skipping Couple." 3" x 4".

"Skating Couple." 2" x 5".

SPORTS AND RECREATION

Football player. 2" x 5".

"Hockey Player." 2" x 4.5".

Baseball catcher. 2" x 6".

"Boxing." 4" x 6".

"Jockey on Horse." 3" x 4".

"Horse Race." L: with horn in hand; R: with sword in hand. 4"
x 5".

Boys on Sleds: L: lying with red cap, 2" x 4"; C: "Lucky
Sledge," 4" x 4"; R: lying with blue cap, 3" x 4".

"Skier." 6" x 5".

"Swim Queen." 3" x 8".

"Skier." 4" x 6.5".

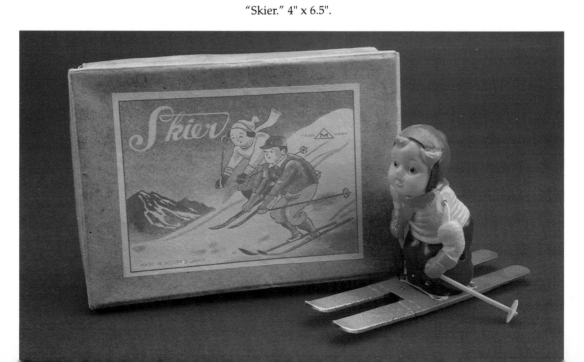

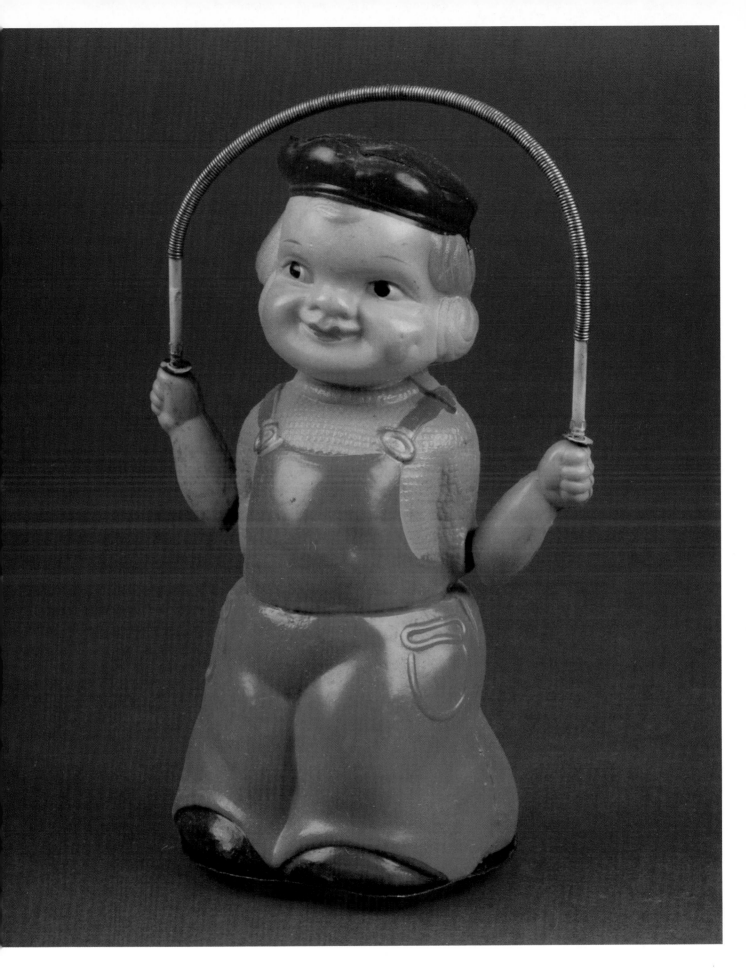

Dutch boy skipping rope. 2" x 5".

"Happy Life." 6" x 9".

"Rowing Boat." 4" x 6".

88

Musicians

Boy beating drum. 2" x 6".

Drummer, lithographed tin body. 2" x 5".

"Jeannie" with violin. 3" x 6".

Musicians, drummer and sax player with parasol. 2" x 7".

"Xylophone Player." 3" x 6".

90

BABIES

"Frightened Boy." 5" x 6".

"Naughty Dog." 5" x 6".

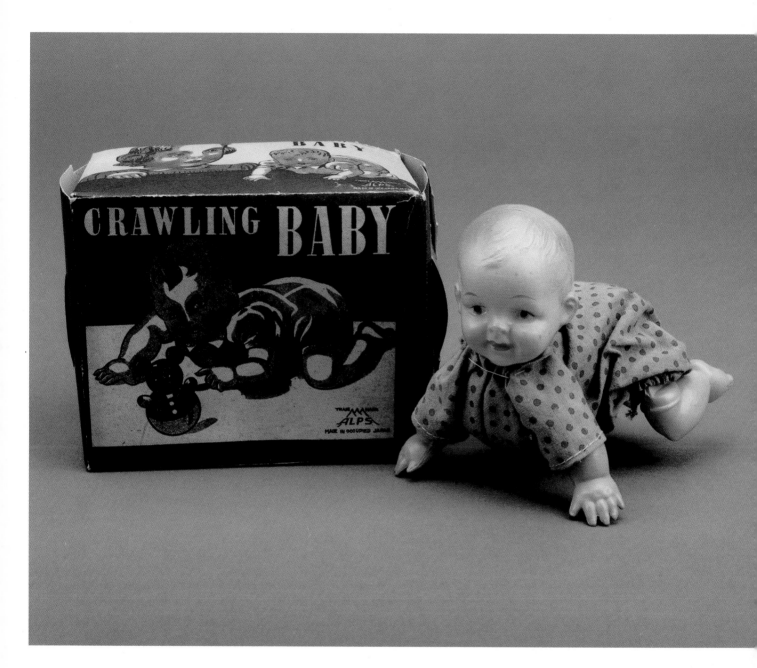

"Crawling Baby." 4" x 5".

"Tumbling Dolls," available in the two colors shown. 2" x 6".

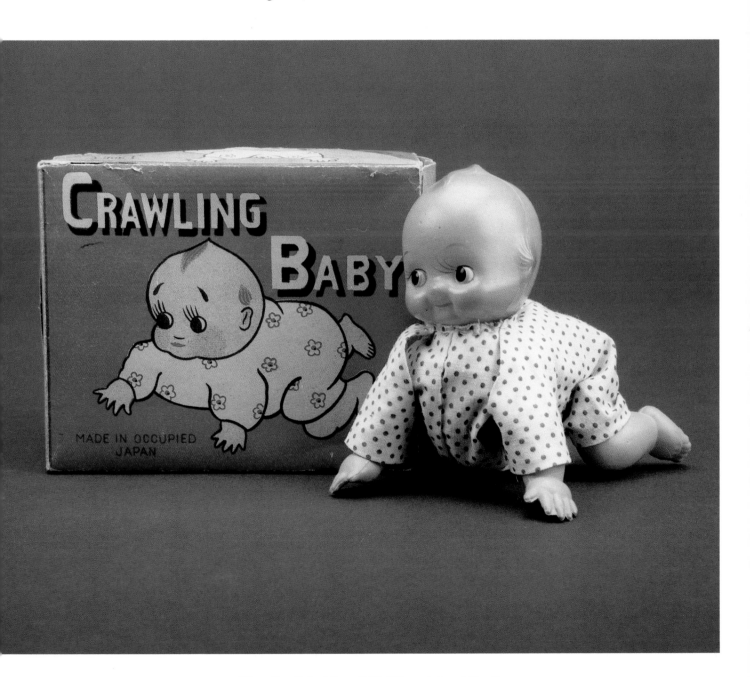

"Crawling Baby," Rose O'Neil Kewpie type! 4" x 4".

"Toddler." 2" x 6".

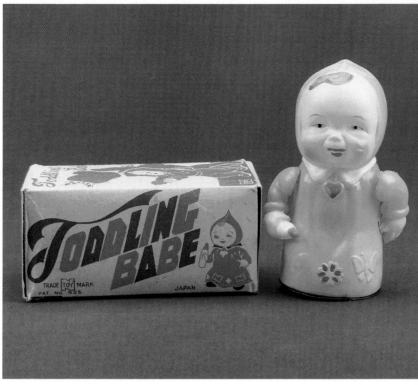

"Toddling babe." 2" x 4".

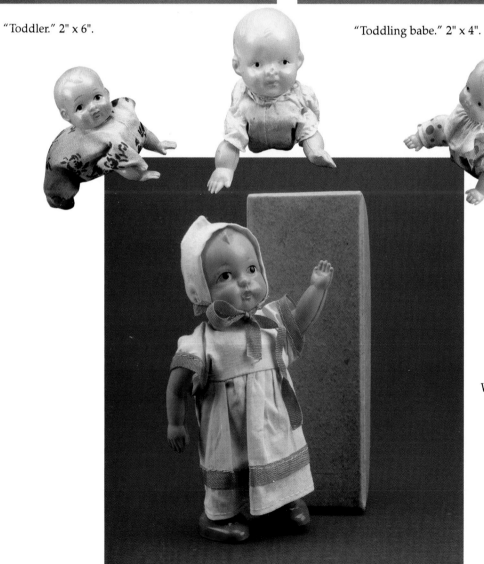

Miscellaneous infants.
Small: 2" x 3"; large, 3" x 4".

Walking toddler. 2" x 6".

"Walking Little Lady." 2" x 7".

"Dandy Girl." 2" x 5".

"Three Dancing Girls." 4" x 5".

"Circus Tricycle" and "Teddy's Cycle." 5" x 5".

Small boy on tricycle. 3" x 3".

"Darling Cycle." 4" x 6".

"Scooter." 5" x 3".

MILITARY

"Sharp Shooter." 3" x 7".

"Kneeling Soldier." 2" x 6".

WORKERS

"Mr. Salesman." 3" x 3".

"Ice Cream Vendor." 4" x 4".

Two cooks: "Cheery" cook," 2" x 4"; "Cheery Cooky," 2" x 4.5".

"Blacksmith," all lithographed tin. 6" x 6". Y.

"News-boy." 2" x 6".

"Billy Boy" porter. 4" x 3".

MISCELLANEOUS

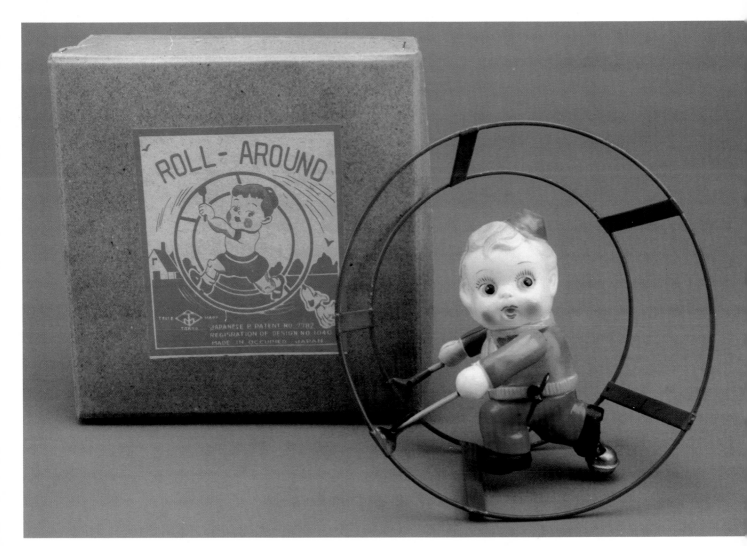

Boy "Roll-Around." 6.5" dia.

Girl "Roll Around." 6.5" dia.

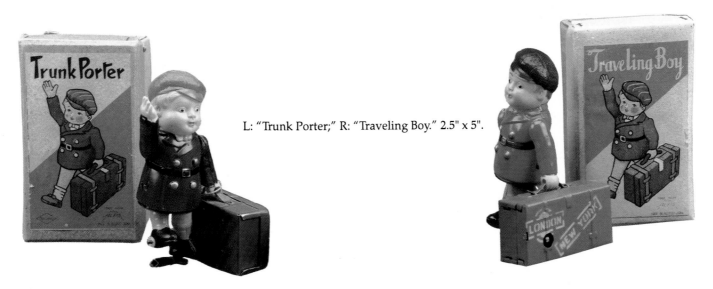

L: "Trunk Porter;" R: "Traveling Boy." 2.5" x 5".

"Drunker." 2" x 6".

"Drunker." 2" x 5.5".

"Happy Tar." 2" x 6".

CHAPTER FOUR CHARACTERS

"Running Mickey on Pluto." 4" x 6".

"Mickey Trapeze." 6" x 8".

"Mickey Race Car" and "Donald Race Car." 2" x 3".

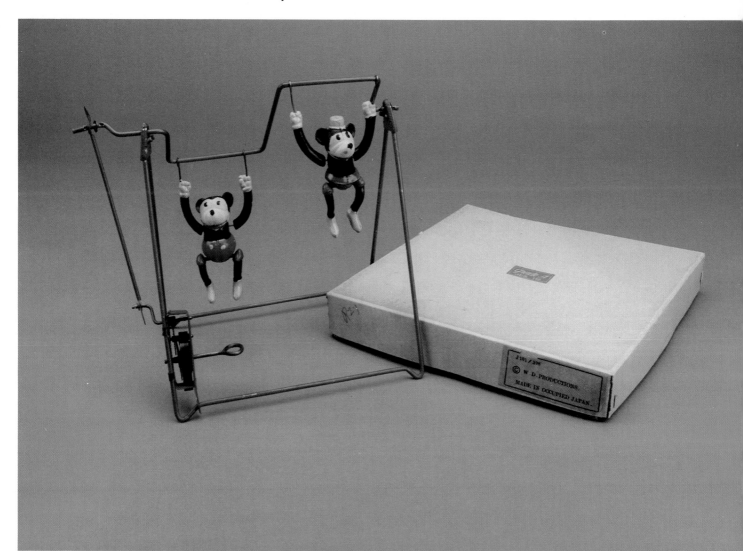

Mickey & Minnie on trapeze 9" x 9".

Donald "long-billed" Duck whirligig. 3" x 7.5".

Mickey Whirligig. 3" x 7.5".

Harold Lloyd, unlicensed. Lithographed tin. 2" x 6".

"Jolly," unlicensed Wimpy. 2" x 6".

"Groggy," Dagwood? 5" x 6".

"Kitten," Felix? 2" x 5.5".

"Egghead" in barrel. 2" x 4".

Fred Astaire? 3" x 9".

"Phineas T. Bluster" from Howdy Doody, 6" x 14".

"Merry & Little Lamb." 4" x 5.5"

CHAPTER FIVE   CIRCUS TOYS

Ringmaster on pony. 4" x 5".

"Circus Parade." 8" x 5".

Clown with parasol. 2" x 6".

Celluloid clown with rattles. 2" x 6".

"Tumbling Pierrot." 3" x 5".

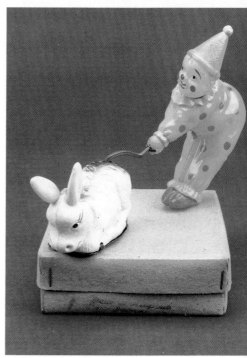

Clown with donkey. 7" x 5".

"Jolly Pig." 7" x 5".

"Fancy Dan the Juggling Man." 2" x 6".

Drunken clown. 2" x 4".

"Mechanical Toy Juggler." 2" x 5".

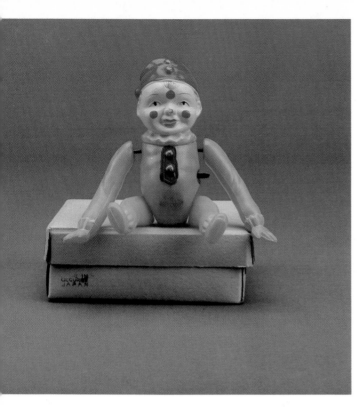

Flip-over clown. 4" x 5".

"Hand Standing Pierrot." 3" x 8".

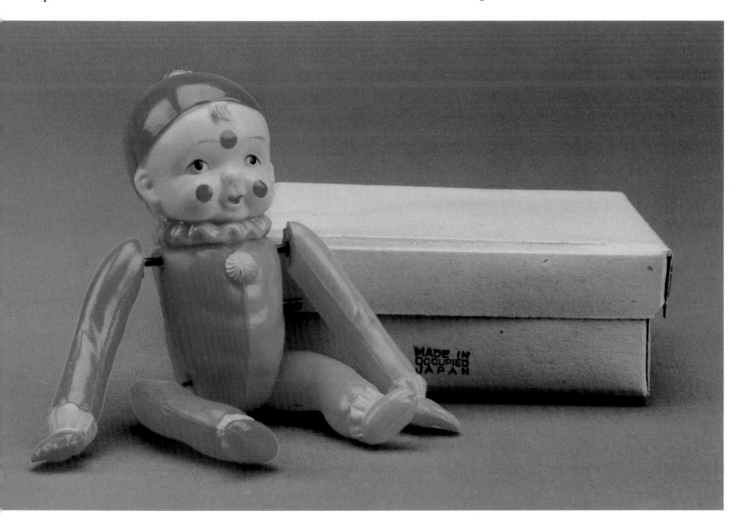

Flip-over clown." 4" x 5".

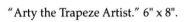
"Arty the Trapeze Artist." 6" x 8".

"Funny Man." 2" x 6".

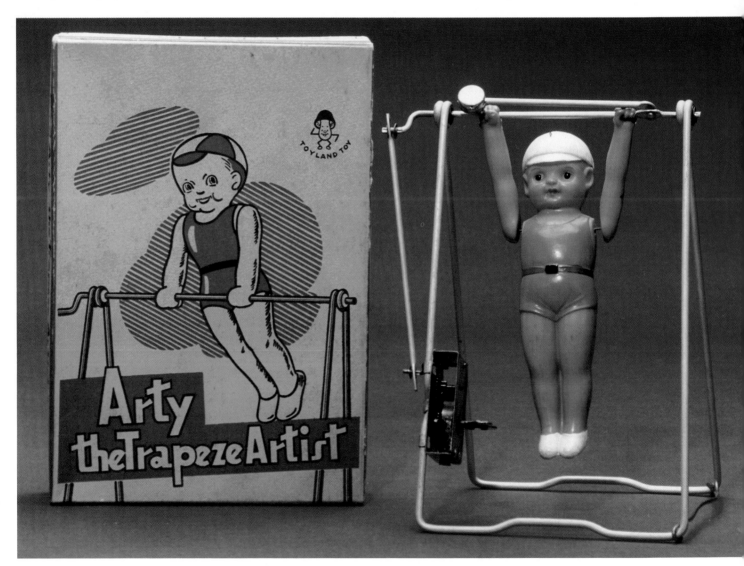
Large "Arty the Trapeze Artist." 7" x 9".

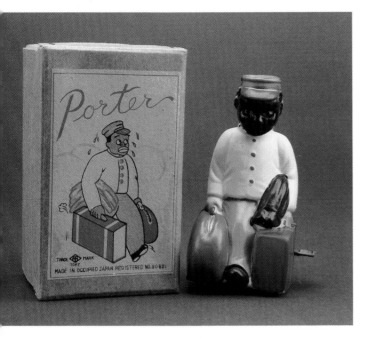

"Porter." 4" x 6". Y.

"Baby Porter." 3.5" tall.

Two black "Tap Dancers," "Lenox Ave. and 125th St" and
"Hollywood and Vine." 3" x 9". Y.

"Cook Boy." 2" x 4.5".

"Poor Pete." 5" x 6".

Black baby. 4" x 4".

Boy on alligator. 2" x 11".

L: "Hawaiian Dancers" (black); R: "Dance Hawaiian" (yellow). 3" x 6".

"Jumbo the Angry Elephant" with native. 5" x 4".

CHAPTER SEVEN
CHRISTMAS AND
EASTER TOYS

"Santa Claus" nodder. 2" x 3.5".

"Walk Santa Claus" with lead legs. 2" x 7".

Santa Claus nodder with bell. 2" x 4.5".

Santa Claus non-moving. 2" x 6".

Santa rattle.

Santa Claus with four deer. 2.5" x 7".

"Santa Claus on Sled." 4" x 8".

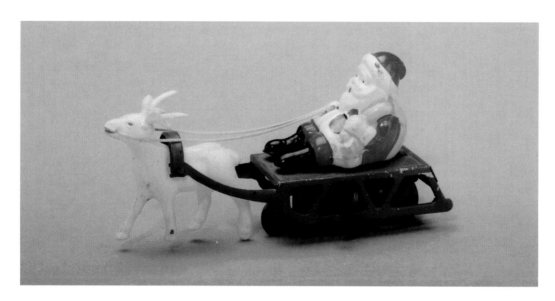

Santa. 1" x 4.5".

"Merry Christmas." 4" x 8".

Santa on sled. 1.5" x 5".

"Santa Express," with lead antlers. 4.5" x 9".

"Easter Parade." 4" x 8".

123

"Mechanical Easter Toy." 4" x 8".

"Easter on Parade." 4" x 8".

"Atomic Robot Man." 2" x 4".

"Lucky Go Round," space kaleidoscope. 3" x 8".

Chapter Eight
Miscellaneous
Mechanical Toys

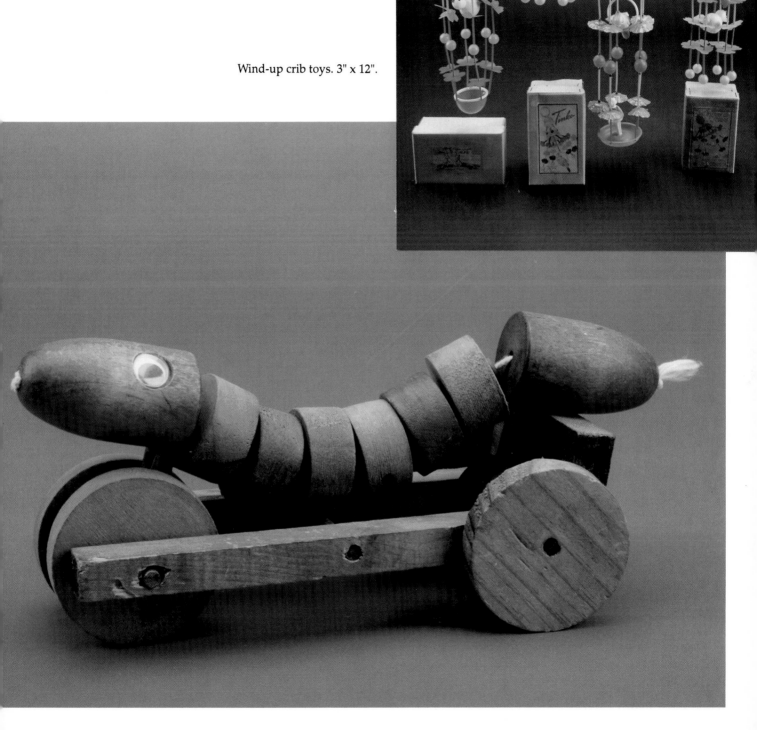

Wind-up crib toys. 3" x 12".

Wooden pull toy, worm. 4" X 9".

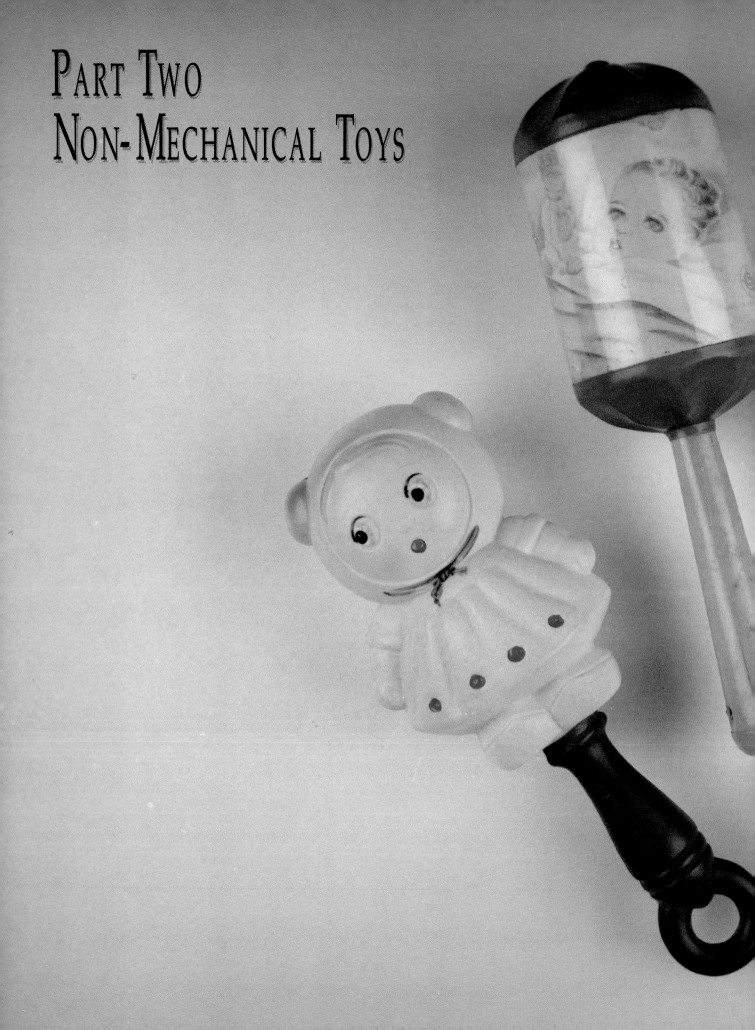

PART TWO
NON-MECHANICAL TOYS

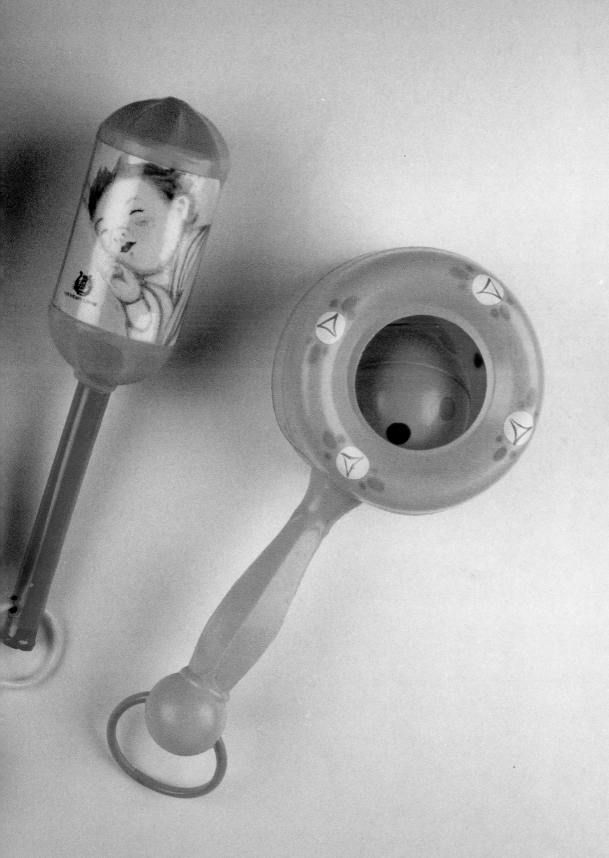

Assorted stuffed animals. 4-9".

Assorted stuffed animals. 4-5".

CHAPTER TEN BISQUE DOLLS

Twin black jointed baby dolls. 4"

4" black jointed baby doll.

Black jointed twin dolls. 3".

Assorted dolls.

Three bisque dolls. 3-5".

Assorted jointed bisque dolls. 4-8".

131

Set of five bisque dolls.

Set of five jointed bisque dolls in original box, (Dionne Quints?)

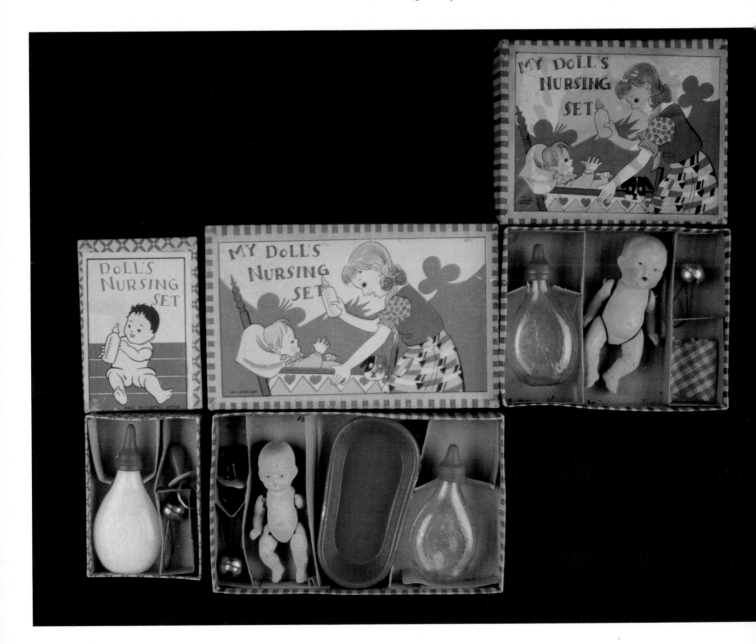

Assorted dolls nursing sets in original boxes.

CHAPTER ELEVEN CELLULOID TOYS

Assorted celluloid dolls. 4".

3-5" Kewpie dolls.

Two 12" celluloid dolls.

6" assorted black dolls.

134

3-6" assorted dolls.

Three 10" celluloid dolls.

7-9" celluloid dolls.

Bride and groom, 4".

Football players, 3" and 10".

3-5" celluloid dolls.

Football player, 5".

Baby squeak toys. 4-5".

Baby rattles and roly-poly. 4-6".

3-7" celluloid bears.

4-6" celluloid animals.

Assorted birds, 2-12".

Assorted animals, 3-9".

139

3-5" assorted boats.

3-6" celluloid animal nodders.

Assorted baby rattles, 4-8".

140

Chapter Twelve Character Toys

Unlicensed Mickey, Donald, and Jiminy Crickets ornaments. 3-6".

5" Donald Duck rattle.

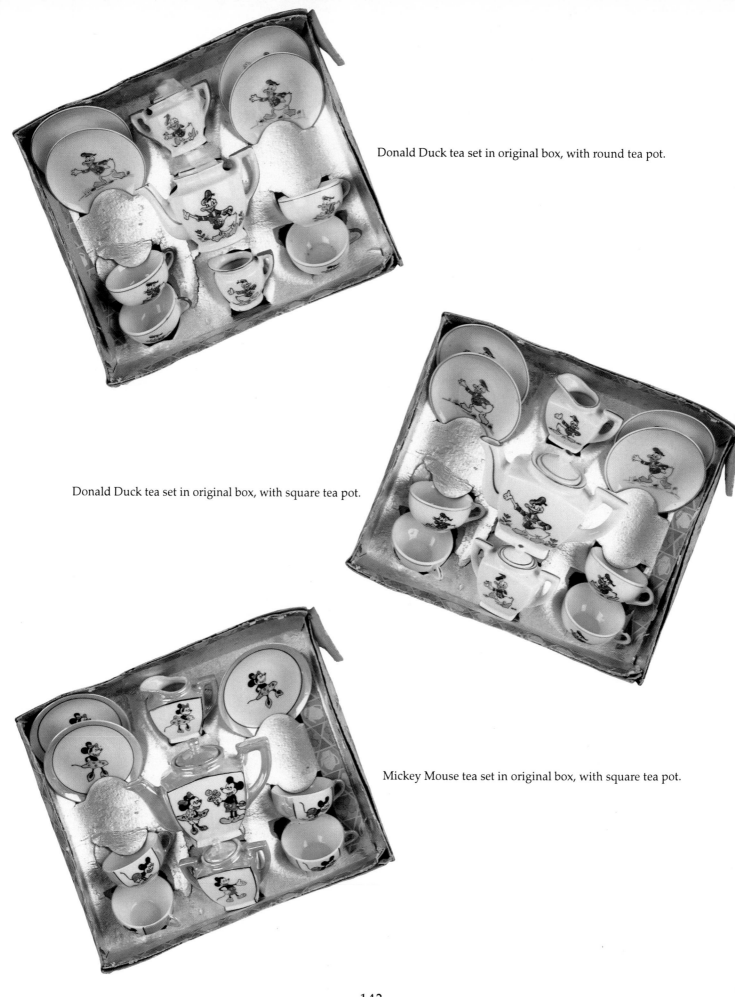

Donald Duck tea set in original box, with round tea pot.

Donald Duck tea set in original box, with square tea pot.

Mickey Mouse tea set in original box, with square tea pot.

CHAPTER THIRTEEN MISCELLANEOUS NON-MECHANICAL TOYS

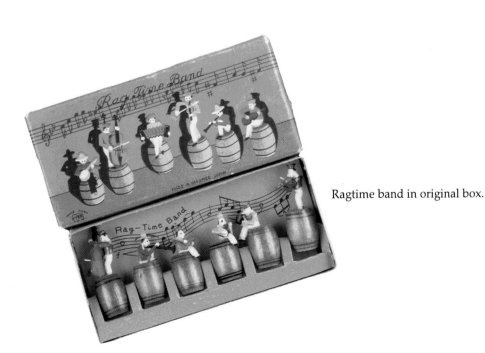

Four sets of leaded figures.

Ragtime band in original box.

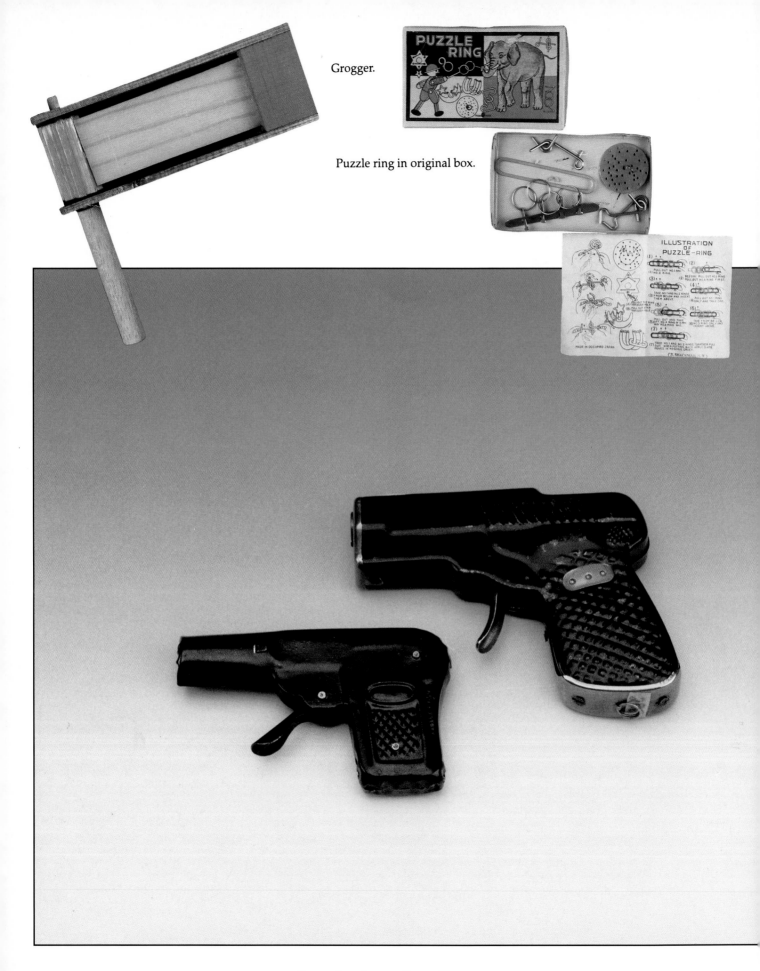

Grogger.

Puzzle ring in original box.

L: metal squirt gun; R; sparkling gun.

Novelty camera with pop-out worm.

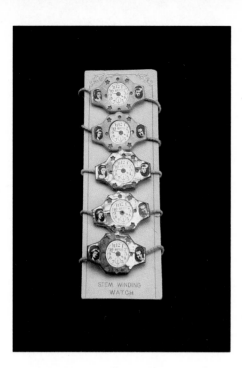

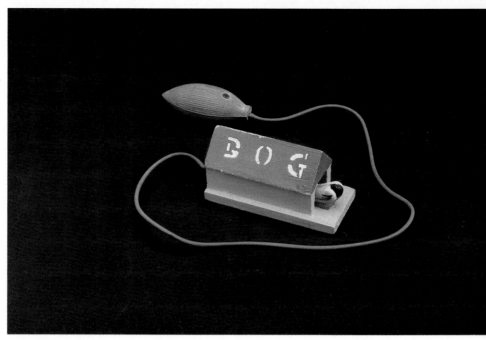

Movie star toy wristwatches (Gary Cooper, Linda Darnell, etc.)

Doghouse squeeze toy. 2" x 4".

Battery operated owl lantern. 5" tall.

Squeaking clown. 9".

Mikky Phone, wind up gramophone. Folded size, 4" x 6".

APPENDIX MANUFACTURER'S CATALOGS

MASUDAYA'S MECHANICAL TOYS CATALOGUE

TRADE **MARK**

TOKYO

MASUDAYA TOYS CO., LTD.

Main Office; 1, 2, 3, Kuramae Daito-Ku Tokyo

TEL: (84) 1620 1, 1631-2

Cable Address: "MASUTOKU" Tokyo

Samples : Assorted if desired, may be sent by parcel port on receipt of your remittance of $100.00 U.S. Cy., or its equivalent. If Air Express desired, please make your letter as follows: "Masudaya Toys Co., Ltd. Please send us your samples via Pan American Air-ways or Northwest Airlines or B-O-A-C, on the pay collect system."

Prices : Our price is fixed by U.S. Cy., base F.O.B. Japanese Port, and will be changed without notice.

The newest items : WELCOME YOUR SUGGESTION & IDEA.

Celluloid
Doll, Animal, Fish & Boat
every size.

Mechanical
1635 Merry-go-round 11×3½×3
Sweet music with chime, one wind
loops 10 minutes
15 doz. 18 cft.
Net 116 lbs. Gross 160 lbs.
each in box.
Monthly Production 100 gross

Terms & Conditions : As per the Japanese Board of Trade form JX 10 Buyer-Supplier Sales Contract.

(1) Irrevocable Letter of Credit, in the amount equal to 110% of the estimated total purchase price is to be established in favor of Masudaya Toys Co., Ltd.

(2) Insurance & W.R. are to be covered by buyers.

Bank References :
The Bank of Tokyo Ltd., Tokyo
The National City Bank of New York, Tokyo.
The Chase National Bank of the City of New York, Tokyo.

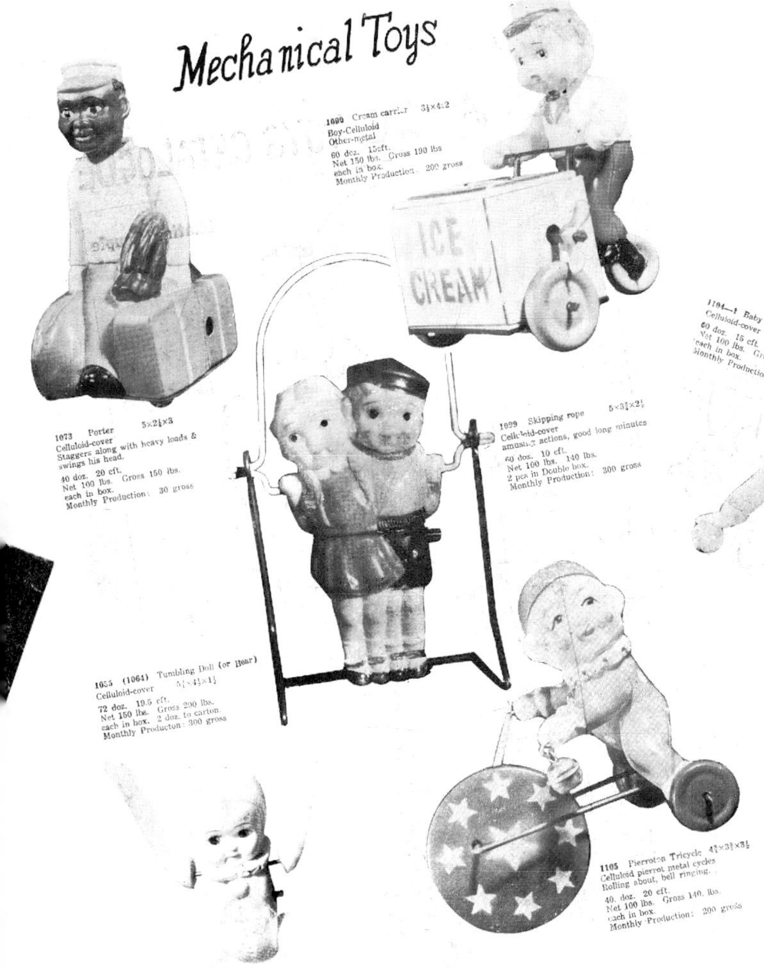

Mechanical Toys

1080 Cream carrier 3¼×4×2
Boy-Celluloid
Other-metal
60 doz. 15cft.
Net 150 lbs. Gross 190 lbs
each in box.
Monthly Production 200 gross

1073 Porter 5×2¼×3
Celluloid-cover
Staggers along with heavy loads & swings his head.
40 doz. 20 cft.
Net 100 lbs. Gross 150 lbs.
each in box.
Monthly Production: 30 gross

1099 Skipping rope 5×3¾×2½
Celluloid-cover
amusing actions, good long minutes
60 doz. 10 cft.
Net 100 lbs. 140 lbs.
2 pcs in Double box.
Monthly Production: 300 gross

1055 (1064) Tumbling Doll (or Bear) 5¼×4½×1½
Celluloid-cover
72 doz. 19.5 cft.
Net 150 lbs. Gross 200 lbs.
each in box. 2 doz. to carton.
Monthly Producton: 300 gross

1105 Pierroton Tricycle 4¾×3¾×3½
Celluloid pierrot metal cycles
Rolling about, bell ringing.
40. doz. 20 cft.
Net 100 lbs. Gross 140. lbs.
each in box.
Monthly Production: 200 gross

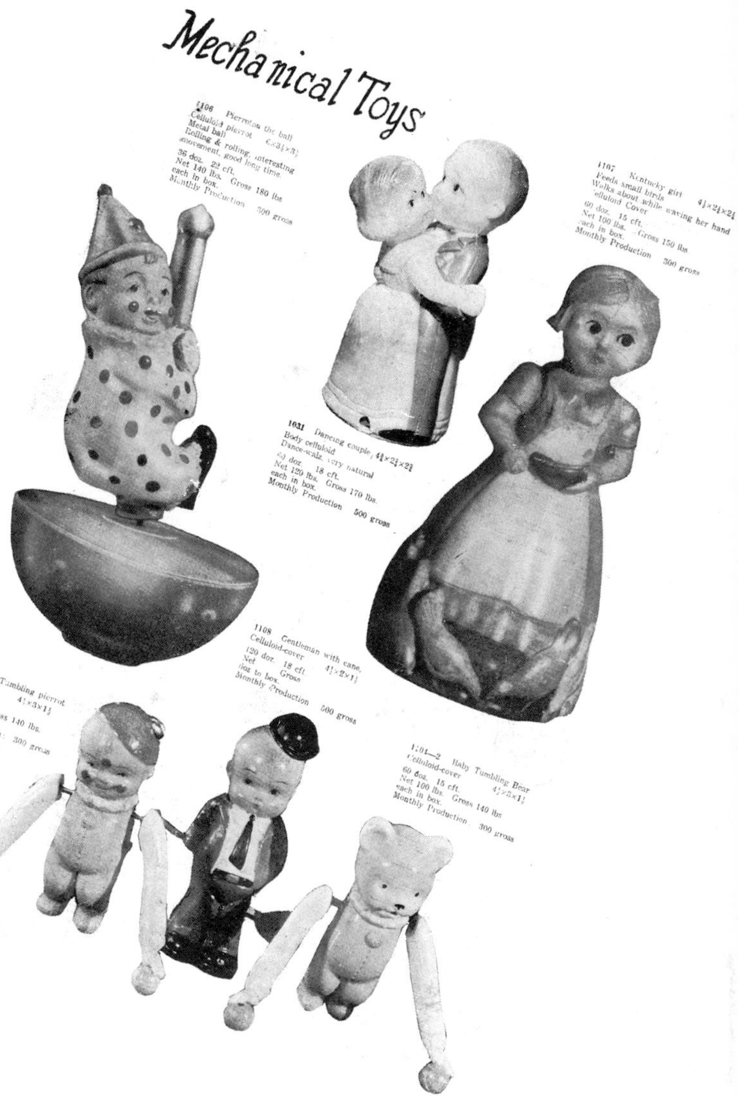

Mechanical Toys

1106 Pierroton the ball
Celluloid pierrot 6×3¾×3¾
Metal ball
Rolling & rolling, interesting movement, good long time.
36 doz. 22 cft.
Net 140 lbs. Gross 180 lbs
each in box.
Monthly Production 300 gross

1107 Kentucky girl 4¼×2¼×2¼
Feeds small birds
Walks about while waving her hand
Celluloid Cover
60 doz. 15 cft.
Net 100 lbs. Gross 150 lbs.
each in box.
Monthly Production 300 gross

1031 Dancing couple, 4¾×2¼×2¾
Body celluloid
Dance-walz, very natural
60 doz. 18 cft.
Net 120 lbs. Gross 170 lbs.
each in box.
Monthly Production 500 gross

1108 Gentleman with cane,
Celluloid-cover 4½×2×1½
120 doz. 18 cft.
Net Gross
doz to box.
Monthly Production 500 gross

1104—1 Baby Tumbling pierrot
Celluloid-cover 4½×3×1½
60 doz. 15 cft.
Net 100 lbs. Gross 140 lbs.
each in box.
Monthly Production: 300 gross

1104—2 Baby Tumbling Bear
Celluloid-cover 4½×3×1½
60 doz. 15 cft.
Net 100 lbs. Gross 140 lbs
each in box.
Monthly Production 300 gross

Mechanical Toys

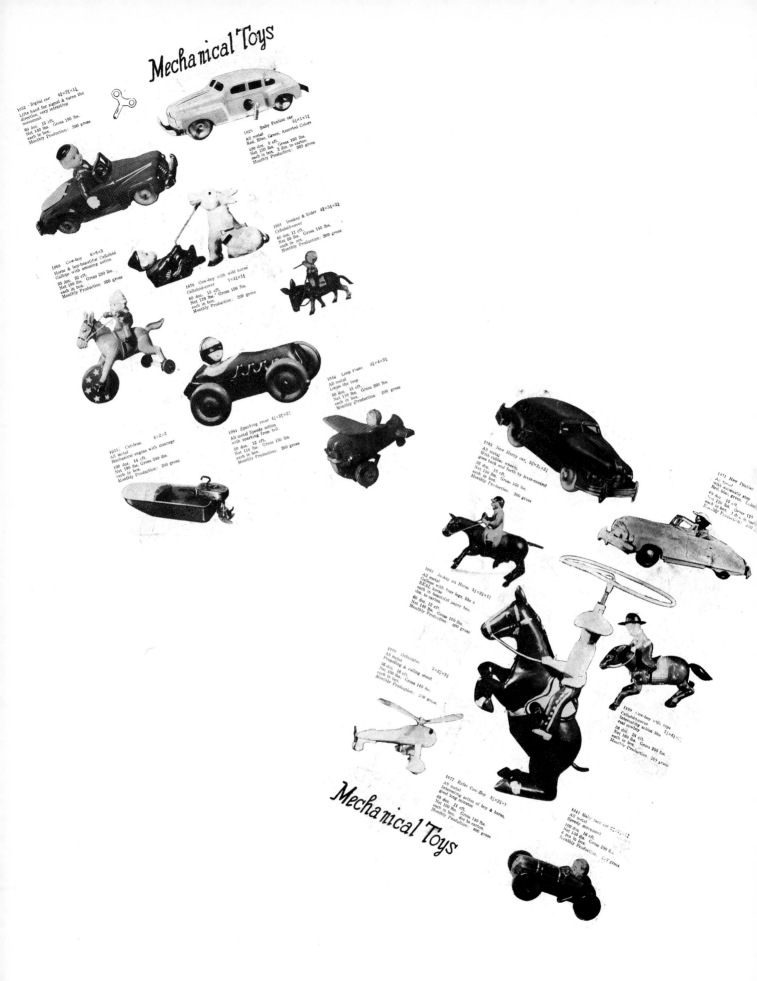

1052   Signal car   4½×2½×1½
Lifts hand for signal & turns the
direction, very intresting
movement.
60 doz.   15 cft.
Net 140 lbs.   Gross 190 lbs.
each in box.
Monthly Production: 300 gross

1025   Baby Pontiac car   3½×1×1½
All metal
Red, Blue, Green, Assorted Colors
130 doz.   9 cft.   Gross 150 lbs.
Net 130 lbs.   2 dns to carton.
each in box.   2 dns. to carton.
Monthly Production: 500 gross

1069   Cow-boy   6×6×3
Horse & boy-beautiful Celluloid
Gallops with amusing action
25 doz.   22 cft.
Net 180 lbs.   Gross 280 lbs.
each in box.
Monthly Production: 300 gross

1078   Cow-boy with wild horse
Celluloid-cover
60 doz.   15 cft.
Net 110 lbs.   Gross 150 lbs.
each in box.
Monthly Production: 200 gross

1091   Donkey & Rider   4½×3½×2½
Celluloid-cover
60 doz.   11 cft.
Net 90 lbs.   Gross 140 lbs.
each in box.
Monthly Production: 200 gross

1048   Loop Plane   3½×4×2½
All metal
Loops the loop
60 doz.   15 cft.
Net 110 lbs.   Gross 260 lbs.
each in box.
Monthly Production: 200 gross

1091   Sparking racer   4½×2½×2½
All metal Speedy action
with sparking from tail.
50 doz.   12 cft.   Gross 150 lbs.
Net 110 lbs.
each in box.
Monthly Production: 200 gross

1015   Cut-boat   6×2×2
All metal
Mechanical engine with steerage
100 doz.   14 cft.
Net 190 lbs.   Gross 240 lbs.
each in box.
Monthly Production: 200 gross

1182   New Merry car   5½×2½×2½
All metal
With rubber wheels,
goes back and forth by lever-control
36 doz.   15 cft.
Net 130 lbs.   Gross 100 lbs.
each in box.
Monthly Production: 300 gross

1071   New Pontiac
All metal
With automatic stop
Red, blue, green
60 doz.   10 cft.   Gross 170
Net 120 lbs.   1 d n. in car
each in box.
Monthly Production: 820

1085   Jockey on Horse   5½×3½×1½
All metal
Gallops with four legs, like a
REAL horse in beautiful paper box.
dns. to carton
60 doz   12 cft
Net 140 lbs   Gross 180 lbs.
Monthly Production: 400 gross

1110   Helicopter   7×3½×5½
All metal
Propelling & rolling about
36 doz.   18 cft.
Net 100 lbs.   Gross 140 lbs.
each in box.
Monthly Production: 230 gross

1116   Cow-boy with rope
Celluloid-cover   7×4½×2
Interesting action like
real cowboy
36 doz.   24 cft.
Net 160 lbs.   Gross 200 lbs.
each in box.
Monthly Production: 200 gross

1077   Rider Cow-Boy   3½×3½×1
All metal
Interesting action of boy & horse,
good long 3 minute
60 doz.   11 cft.
Net 100 lbs.   Gross 140 lbs.
each in box.   dns to carton.
Monthly Production: 400 gross

1041   Baby race car   3½×1½×1½
All metal
Speedy movement
100 doz.   16 cft.
Net 130 lbs.   Gross 200 lbs.
each in box.
Monthly Production: 237 gross

Mechanical Toys

1915 Strolling Duck 4×4½×3
Novel toy Celluloid-cover
Beautiful Celluloid-cover
Walks about with ducks.
40 doz. 12 cft. Gross 190 lbs.
Net 15 lbs.   1 doz. in carton.
each in box.
Monthly Production: 200 gross

1005 O.U. Dog 5½×3×1½
Cover-celluloid
Moves head & wags tail
60 doz. 14.5 cft. Gross 140 lbs.
Net 100 lbs.
each in box.
Monthly Production: 200 gross

1659 Shoes with dog 6½×2½
Dog-Celluloid
Shoe-Metal
60 doz. 12 cft. Gross 140 lbs.
Net 100 lbs. doz. in carton.
each in box.
Monthly Production: 300 gross

1112 Walking Bear 5½×3½×2
Cover-celluloid
Walking on four legs, like a real
bear
40 doz. 15 cft.
Net 170 lbs. Gross 210 lbs.
Monthly Production: 200 gross

1074 Elephant 3½×2½×1½
All metal
Movable legs, can stand by
hind legs
100 doz. 13 cft. Gross 180 lbs.
Net 140 lbs. doz. to carton.
each in box.
Monthly Production: 300 gross

Mechanical Toys

1057 Monkey with fruit
All metal
Three actions:
Moves lower jaw, lifts his hands
and wags his tail.
60 doz. 15 cft.
Net 100 lbs. Gross 140 lbs.
each in box.
Monthly Production 300 gross

1056 Pop Monk 5×3×2½
Cover-Celluloid
Makes up with comb & mirror
tail moving at the same time
40 doz. 12 cft.
Net 80 lbs. Gross 150 lbs.
each in box.
Monthly Production: 300 gross

107 Rolls the Monk 5×5×2½
Runs on hoop, good long minutes
30 doz. 18 cft.
Net 80 lbs. Gross 120 lbs
each in box.
Monthly Production: 300 gross

Mechanical Toys

1056 Tumbling Monkey 4×3½×1
Celluloid-cover
120 doz. 14 cft.
Net 90 lbs. Gross 130 lbs.
each in box.   2 doz. in carton.
Monthly Production: 300 gross

1113 Monkey & Bee 4×2½×2
Celluloid-cover
60 doz. 12 cft.
Net 80 lbs. Gross 120 lbs.
each in box.
Monthly Production: 300 gross

153

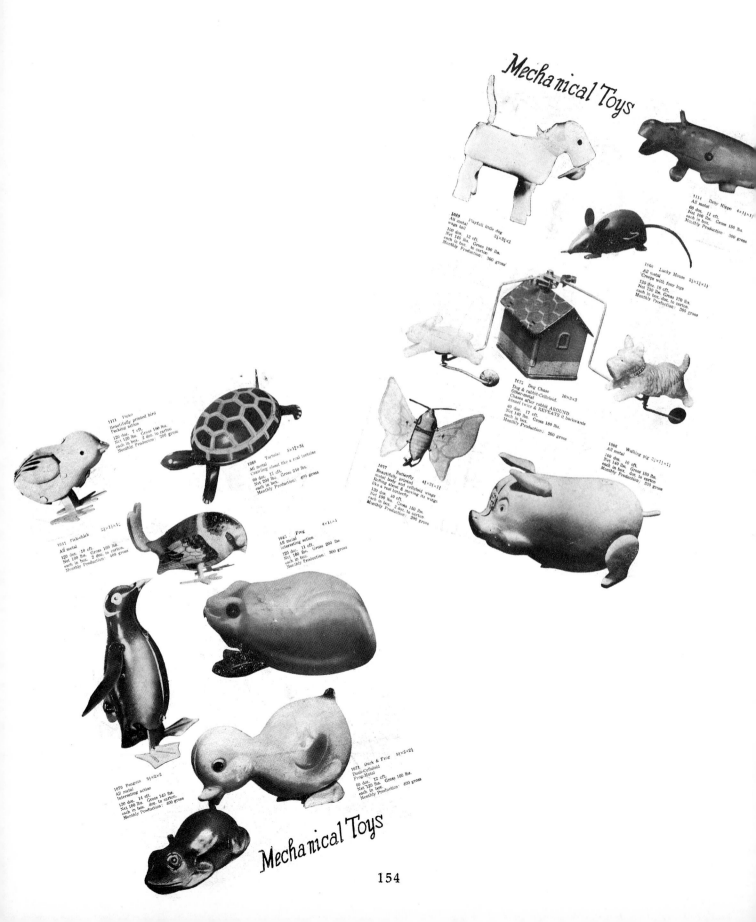

Mechanical Toys

1089 Playfull little dog $3\frac{1}{4}\times3\frac{1}{4}\times1$
All metal
wags tail
100 doz. 12 cft.
Net 140 lbs. Gross 180 lbs.
each in box. to carton
Monthly Production: 300 gross

1114 Baby Hippo $4\times1\frac{1}{4}\times1\frac{1}{2}$
All metal
60 doz. 11 cft.
Net 100 lbs. Gross 150 lbs.
each in box.
Monthly Production 300 gross

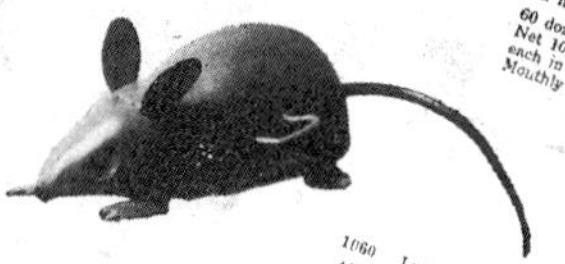

1060 Lucky Mouse $3\frac{1}{4}\times1\frac{1}{4}\times1\frac{1}{4}$
All metal
Creeps with four legs
120 doz. 10 cft.
Net 130 lbs. Gross 170 lbs.
each in box. doz. to carton.
Monthly Production: 300 gross

1075 Dog Chase $10\times3\times3$
Dog & rabbit-Celluloid.
Other-metal
Chases after rabbit AROUND
kennel twice & REPEATS it backwards
40 doz. 17 cft.
Net 140 lbs. Gross 180 lbs.
each in box.
Monthly Production: 200 gross

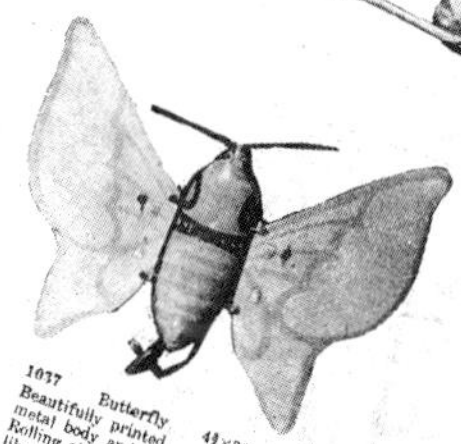

1037 Butterfly $4\frac{3}{4}\times3\frac{3}{4}\times1\frac{3}{4}$
Beautifully printed
metal body and celluloid wings
Rolling about & moving its wings
like a real butterfly
120 doz. 10 cft.
Net 100 lbs. Gross 150 lbs.
each in box. 2 doz. to carton
Monthly Production: 200 gross

1086 Walking pig $3\frac{1}{4}\times1\frac{1}{4}\times1\frac{1}{4}$
All metal
100 doz. 10 cft.
Net 140 lbs. Gross 180 lbs.
each in box. doz. to carton
Monthly Production: 300 gross

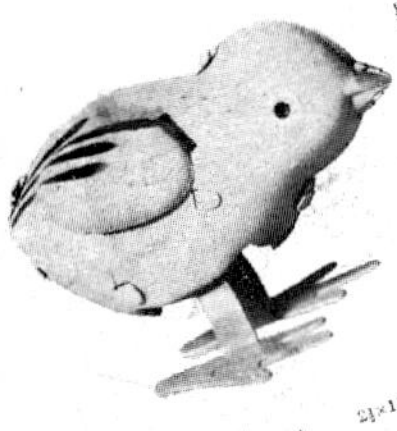

1111 Picko
Beautifully printed bird
Pecking action
120 doz. 7 cft.
Net 120 lbs. Gross 180 lbs.
each in box. 2 doz. to carton
Monthly Producion: 200 gross

1068 Tortoise $5\times1\frac{3}{4}\times3\frac{1}{4}$
All metal
Crawling about like a real tortoise
60 doz. 17 cft.
Net 200 lbs. Gross 240 lbs.
each in box.
Monthly Production: 400 gross

1031 Pick-chick $2\frac{1}{4}\times1\frac{1}{4}\times1\frac{1}{4}$
All metal
120 doz. 10 cft.
Net 100 lbs. Gross 160 lbs.
each in box. 2 doz. in carton.
Monthly Production: 400 gross

1057 Frog $4\times1\frac{1}{4}\times1$
All metal
interesting action
120 doz. 11 cft.
Net 160 lbs. Gross 200 lbs.
each in box.
Monthly Production: 300 gross

1070 Penguin $3\frac{1}{4}\times2\times2$
All metal
Interesting action
120 doz. 14 cft.
Net 100 lbs. Gross 140 lbs.
each in box. doz. to carton.
Monthly Production: 400 gross

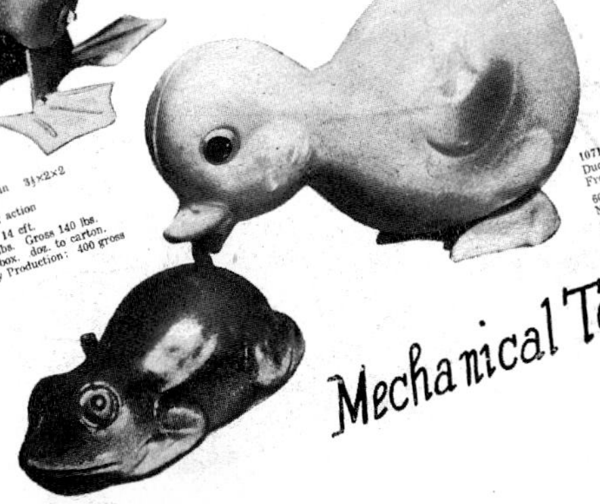

1071 Duck & Frog $5\frac{3}{4}\times2\times2\frac{1}{4}$
Duck-Celluloid
Frog-Metal
60 doz. 12 cft.
Net 120 lbs. Gross 160 lbs.
each in box.
Monthly Production: 400 gross

Mechanical Toys

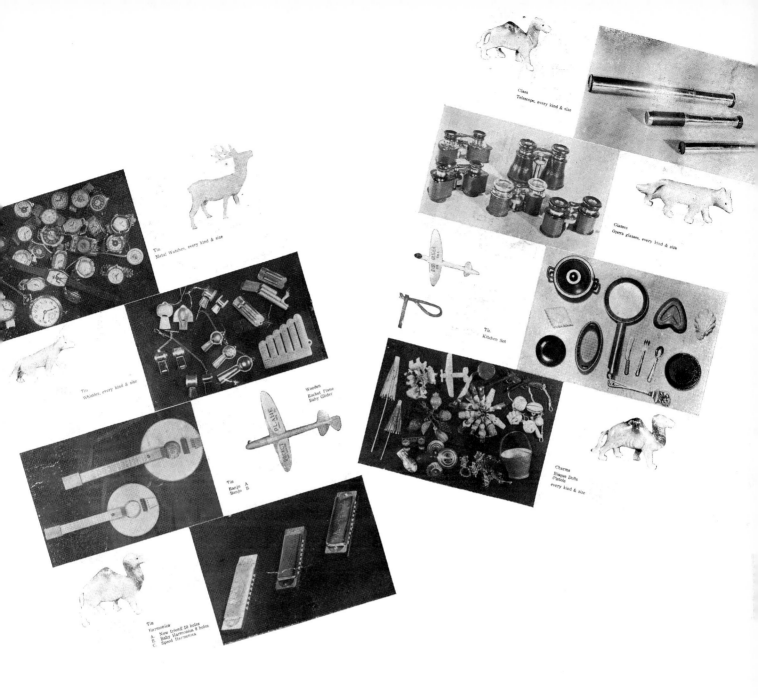

Glass
Telescope, every kind & size

Tin
Metal Watches, every kind & size

Glasses
Opera glasses, every kind & size

Tin
Kitchen Set

Tin
Whistles, every kind & size

Wooden
Rocket Plane
Baby Glider

Charms
Bisque Dolls
/Pistols
every kind & size

Tin
Banjo A
Banjo B

Tin
Harmonica
A. New friend 10 holes
B. Baby Harmonica 8 holes
C. Speed Harmonica

155

PRICE GUIDE

In the value ranges given in the price guide, the lower figure represents a plain toy without its original box and the higher figure represents a toy in near mint condition in its original box. As you will see the presence of the original box profoundly effects the value of the piece.

The prices shown represent what a collector can expect to pay for an object at retail. They may differ somewhat from one geographic location to another, and certainly by the condition of the toy. While the author has many years of experience in the field and has created a guide that is as accurate as possible at the time of publication, he can accept neither credit for any profit the reader may experience by using it, nor blame for any losses.

Position Codes:

These codes refer to the relative position of photograph of an item on the page indicated.

L = left	TR = top right
C = center	CL = center left
R = right	CR = center right
B = bottom	BL = bottom left
T = top	BC = bottom center
TL = top left	BR = bottom right
TC = top center	

PG.	POS.	PRICE	PG.	POS.	PRICE	PG.	POS.	PRICE	PG.	POS.	PRICE
8	C	$100-150	19	B	$50-75	30	B	$90-150	39	B	$30-50
8	T	$75-100	20	TR	$75-150	31	TL	$75-125	40	T	$125-250
8	B	$75-100	20	CL	$150-250	31	TR	$100	40	B	$50-75
9	T	$50-75	20	BL	$75-100	31	B	$400-600	41	T	$100-150
9	B	$50-75	20	CR	$75-100	32	TL	$75-100	41	C	$100-150
10	T	$50-100	21		$200-400	32	TR	$30-50	41	B	$50-75
10	B	$40-60	22	TR	$50-90	32	B	$40-60	42	T	$100-200
11	TL	$60-85	22	B	$60-75	33	BL	$150-200	42	C	$200-400
11	TR	$50-100	22	TL	$40-60	33	BR	$100-150	43	C	$100-200
11	B	$75-150	23	CR	$75-100	33	TR	$25-50	43	B	$75-125
12	TL	$75-125	23	TL	$100-150	33	TL	$40-75	43	T	$75-125
12	CL	$60-85	23	B	$100-150	34	CL	$75-200	44	TL	$50-75
12	B	$75-150	24	T	$100-200	34	TR	$100	44	TR	$100-150
13	T	$50-75	24	B	$150-300	34	B	$100-150	44	B	$300-450
13	C	$50-75	25	TC	$75-100	35	B	$100-150	45	B	$150-300
13	B	$50-100	25	B	$75-100	35	TR	$40-60	45	CL	$30-50
14	B	$100-150	25	TL	$25	35	TL	$100-150	46	TL	$100-150
14	C	$200-350	26	CR	$75-100	36	T	$50-75 (turtle))	46	B	$75-125
14	T	$35-50	26	T	$75-125	36	T	$75-100 (tort.)	47	B	$75-150
15	T	$150-200	26	B	$100-150	36	B	$100-150 (croc.)	47	TL	$200-300
15	B	$150-300	27	TL	$100-150	36	B	$75-100 (kang.)	47	TR	$100-200
16	T	$200-300	27	TR	$75-125	37	TR	$75-100	48	T	$75-150
16	C	$100-250	27	BL	$150-200	37	B	$25-50 ea.	48	C	$125-250
16	B	$100-200	27	BR	$75-100	37	TL	$50-75	49	T	$300-450
17	TL	$100-200	28	B	$75-100	38	T	$100-150	49	B	$200-300
17	TR	$125-250	28	T	$125-150	38	B	$75-150	50	T	$75-150
17	B	$100-150	29	B	$75-125	39	T	$75-150	50	B	$200-350
18	T	$100-150	29	TL	$75-100	39	C	$75-150	51	T	$100-150
18	C	$125-175	29	TR	$75-100				51	B	$500-800
18	B	$75-125	30	T	$75-150				51	C	$100-200
19	T	$75-125	30	C	$75-125				52	B	$150-200

PG.	POS.	PRICE	PG.	POS.	PRICE	PG.	POS.	PRICE	PG.	POS.	PRICE
52	T	$100-200	73	TL	$175-275	94	B	$150-300			$350-400
52	C	$150-250	73	TR	$175-275	95	T	$200-400	120	R	$150-200
53	B	$50-75	73	B	$200-300	95	C	$300-500	121	B	$250-400
53	TR	$75-100	73	C	$200-300	95	B	$500-700	122	T	$100-250
53	TL	$100-200	74		$700-1600	96	C	$100-150 ea.	122	B	$250-350
54	TL	$200-300	75	T	$200-400	97	B	$150-300	123	T	$150-300
54	B	$150-350	75	C	$100-200	98	TL	$100-150	123	C	$400-500
54	TR	$75-150	75	B	$100-200	98	TR	$400-500	123	B	$150-250
55	TL	$250-350	76	TL	$200-400	98	BR	$100-250	124	T	$150-250
55	TR	$200-300	76	TR	$100-150	99	TL	$75-125	124	C	$100-200
55	B	$100-200	76	B	$100-150	99	TR	$150-250	124	BL	$500-1000
56	B	$50-75	77	C	$150-250	99	B	$600-850	124	BR	$300-700
56	TR	$100-200	77	T	$300-400	100	TL	$600-900	125	B	$150-250
56	C	$150-200	77	B	$200-300	100	TR	$150-300	125	TR	$125-250
57	B	$200-300	78	T	$200-300	100	B	$300-400	128		$20-75
57	T	$150-300 ea.	78	BL	$300-400	101	T	$350-450	129		$20-25
58	T	$100-200	78	BR	$400-500	101	BL	$150-250	130	C	$75
58	B	$100-150	79	TL	$100-200	101	BR	$150-250	130	TR	$100 pr.
59	T	$50-75	79	CR	$200-300	102	TL	$200-350	130	B	$75-150
59	B	$100-200	79	B	$250-350	102	TR	$200-350	131	T	$10-40
60	T	$75-125	80	T	$250-350	102	B	$300-400	131	C	$20-40
60	C	$100-200	80	C	$250-350	103	T	Auctioned at	131	B	$20-75
60	B	$100-200	80	BC	$300-500			$9500	132	TL	$50
61	T	$100-200	80	BL	$150-300	103	B	$800-1800	132	TR	$75-150
61	C	$100-200	81	TR	$150-200	104	B	$1500-3000	132	B	$20-50
61	B	$100-200	81	TL	$100-200	104	T	$800-1500 ea.	133	T	$20-40
62	T	$100-150	81	B	$500-800	105	B	$1000-2000	133	B	$20-40
62	C	$100-150	82	TL	$75-150	106		$500-1000	134	B	$50-150
62	B	$100-350	82	TC	$75-150	108	TL	$900-1800	134	T	$25-50 ea.
63	T	$200-400	82	CR	$100-150	108	B	$300-500	135	T	$15-40
63	B	$75-200	82	BC	$75-150	108	TR	$500-800	135	C	$20-40 ea.
63	C	$100-250	82	BR	$200-300	109	B	$500-700	135	B	$20-40 ea.
64	T	$200-300	83	T	$200-400	109	TR	$750-1000	136	B	$10-25 ea.
64	C	$275-400	83	B	$200-300	109	TL	$300-500	136	TL	$40-50
64	B	$200-350	84	TL	$200-300	09	TC	$500-800	136	TR	$10-50 ea.
65	TL	$100-150	84	C	$200-350	110	CL	$400-500	137	TL	$35
65	TC	$200-500	84	CR	$100-200	110	B	$250-400	137	B	$40-90 ea.
65	TR	$150-300	84	BL	$200-350	111		$150-350	137	TR	$35-50 ea.
65	B	$100-125	85	T	$150-250	112	TL	$200-350	138		$10-40 ea.
66	T	$100-200	85	C	$150-200 ea.	112	TR	$200-300	139	T	$20-40 ea.
66	B	$300-400	85	BL	$75-125	112	B	$300-400	139	C	$10-75
67	C	$100-250	85	BC	$100-200	113		$300-500	139	B	$10-50
67	B	$100-250	85	BR	$100-200	114	TL	$150-300	140	C	$15-50 ea.
67	T	$100-200	86	C	$200-350	114	C	$250-350	140	T	$15-25
68	T	$350-600	86	T	$300-500	115	TL	$150-200	140	B	$30-75
68	CL	$100-150	86	B	$250-350	115	B	$150-200	141	T	$25-75 ea.
68	CR	$200-400	87		$100-300	115	TR	$350-500	141	B	$100-150
68	BR	$100-200	88	T	$500-800	116	TR	$400-600	142	T	$300-500
68	BL	$100-200	89	TL	$250-400	116	TL	$200-300	142	C	$300-500
69	TR	$100-250	89	TC	$200-300	116	B	$250-350	142	B	$400-700
69	TL	$100-200	89	B	$500-600	117	TL	$400-700	143	T	$100-150 ea.
69	CL	$100-200	90	TL	$150-300 ea.	117	TR	$300-500	143	B	$35-50
69	C	$100-150	90	B	$150-300	117	B	$500-950	144	TL	$20-25
69	B	$100-200	91	B	$300-500	118	TR	$1000-1500	144	TR	$25-35
70	C	$350-450	91	T	$300-500	118	TL	$400-750	144	B	$30-50
70	B	$400-600	92		$100-200	118	BR	$300-500	145		$20-30
71	T	$300-500	93	T	$100-150	118	CL	$300-500	146	TL	$50-75/card
71	C	$75-150	93	B	$350-500	119	CR	$300-500	146	TR	$25-35
71	B	$350-500	94	TL	$200-400	119	TL	$100-200	146	B	$25-50
72	T	$300-400	94	TR	$100-150	119	BR	$800-2000	147		$100-200
72	C	$400-500	94	C	$75-150	119	BL	$250-350	148		$400-600
72	B	$200-500									

BIBLIOGRAPHY

Archabault, Florence. *Occupied Japan for Collectors.* Atglen, Pennsylvania: Schiffer Publishing, 1992.

Chandler, Ceil. *Occupied Japan-Buyers & Sellers Directory: No. 1.* Houston, Texas: Self-published, 1980

_____. *Made In Occupied Japan.* Houston, Texas: Self-published, 1972.

_____. *Supplement to "Made in Occupied Japan".* Houston, Texas: Self-published, 1973

_____. *Toys & Dolls-Made in Occupied Japan.* Houston, Texas: Self-published, 1973

Etgen, William. *"OJ"-Made in Occupied Japan: Guide Book #1.* Sacramento, California: Self-published, 1969.

Florence, Gene. *The Collectors Encyclopedia of Occupied Japan Collectibles, First in Series.* Paducah, Kentucky: Collector Books, 1976.

_____. *The Collectors Encyclopedia of Occupied Japan Collectibles, Second in Series.* Paducah, Kentucky: Collector Books, 1979.

_____. *The Collectors Encyclopedia of Occupied Japan Collectibles, Third in Series.* Paducah, Kentucky: Collector Books, 1987.

_____. *The Collectors Encyclopedia of Occupied Japan Collectibles, Fourth in Series.* Paducah, Kentucky: Collector Books, 1990.

_____. *The Collectors Encyclopedia of Occupied Japan Collectibles, Fifth in Series.* Paducah, Kentucky: Collector Books, 1992.

Gould, David C., and Donna Creaver Donaldson. *Toys from Occupied Japan with Prices.* Gas City, Indiana: L-W Book Sales, 1993.

Klampkin, Marian. *Made in Occupied Japan: A Collector's Guide.* New York: Crown Publishers Inc., 1976

Sieloff, Judi Ludwig. *Collectibles of Occupied Japan.* Des Moines, Iowa: Wallace Homestead, 1978.

NOTES

NOTES